Corporate Real Estate Asset Management

Strategy and Implementation

Barry P. Haynes and Nick Nunnington

AMSTERDAM • BOSTON • HEIDELBERG • LONDON • NEW YORK
OXFORD • PARIS • SAN DIEGO • SAN FRANCISCO • SINGAPORE
SYDNEY • TOKYO
EG Books is an imprint of Elsevier

EG Books is an imprint of Elsevier
The Boulevard, Langford Lane, Kidlington, Oxford, OX5 1GB, UK
30 Corporate Drive, Suite 400, Burlington, MA 01803, USA

First edition 2010

British Library Cataloguing-in-Publication Data
A catalogue record for this book is available from the British Library

Library of Congress Cataloging-in-Publication Data
A catalog record for this book is available from the Library of Congress

ISBN 978-0-7282-0573-4

For information on all EG Books publications
visit our website at www.elsevierdirect.com

Printed and bound in Great Britain

10 11 12 10 9 8 7 6 5 4 3 2 1

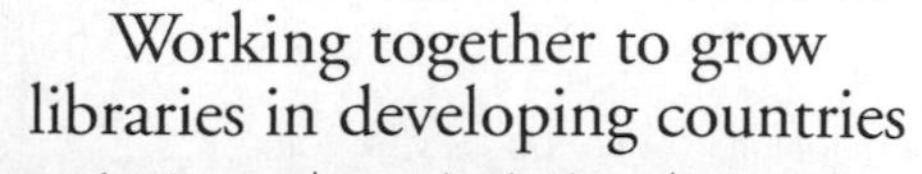

Contents

Preface *ix*
Acknowledgements *xiii*

1 The business environment **1**

Introduction 1
Drivers, issues and trends 1
Significant themes 4
Multi-generational workforce 5
Strategic analysis tools 12
Future scenarios 19
Think the unthinkable 27
Summary checklist 28
References 28

2 Strategic alignment **31**

Introduction 31
An integrated approach: collapsing of boundaries 31
Corporate strategy 36
Adding value 45
CREAM Strategic Alignment Model 52
Summary checklist 54
References 54

3 Strategic real estate procurement options **57**

Introduction 57
The freehold *vs.* leasehold debate 57
The financial argument 62
Getting the right balance of procurement 64
Converting from freehold to leasehold 69
Future implications for procurement options 73
Beyond sale-and-leaseback: strategic alternatives to traditional freehold or leasehold procurement 79

Summary checklist 82
References 83

4 Customer focused procurement options 85

Introduction 85
Changing expectations 85
The financial argument to customer focused management 89
Reducing lease length 92
Code for Leasing Business Premises 94
The RICS Service Charge Code 97
The development of customer focused strategies in the UK 99
Conclusion 104
Summary checklist 105
References 105

5 Performance measurement and benchmarking 107

Introduction 107
Some definitions of benchmarking 107
Benchmarking in practice 110
The 'apples and pears' problem 112
Criticisms of benchmarking 115
Benchmarking in action 1 118
Benchmarking in action 2 121
Benchmarking in action 3 126
Benchmarking in action 4 130
Holistic real estate benchmarking in practice 131
Benchmarking and CREAM strategy 134
The Balanced Scorecard 136
Conclusion 139
Summary checklist 139
References 140

6 Workplace transformation 141

Introduction 141
Evolution of the workplace 141
Changing nature of work 145
Working in the office 147
Working out of the office 151
Working virtually 152
Embedding performance and productivity 153

Conclusion 169
Summary checklist 169
References 170

7 Corporate relocation **173**

Introduction 173
Strategic space planning alignment process 173
Establishing organizational and business demand 175
Briefing process 187
Building and workplace supply 188
Optimum alignment 195
Summary checklist 198
References 198

8 Sustainability, corporate social responsibility and corporate real estate **201**

Introduction 201
Definitions 201
The business benefits of corporate social responsibility (CSR) 203
The professional perspective 206
Sustainability and commercial property 207
Drivers of sustainability and how they impact upon corporate real estate 208
BREEAM 212
Energy Performance Certificates 215
The measurability of sustainability 216
Putting sustainability into practice 219
Green leases 221
The carbon reduction commitment 227
Conclusions 228
Summary checklist 229
References 230
Further Reading 231

CASE STUDIES **233**

Case study 1: Customer focused real estate: The Bruntwood way 233
Case study 2: Benchmarking: Actium Consult – a fresh approach to real estate 243
Case study 3: Workplace transformation: Nokia connecting people 253
Case study 4: Headquarter reconfiguration: The Hong Kong Jockey Club 265
Danny S.S. Then

Case study 5: Sustainability: Amazon Court, Prague – a win–win solution 277

Index *289*

Preface

In this book we provide tools and techniques that can be used to help develop Corporate Real Estate Asset Management (CREAM) strategies. In addition to strategy development many practical examples and case studies relating to CREAM have been used to demonstrate the implementation of strategy into practice. The book is aimed (as in the title) to demonstrate how strategy can be successfully implemented to create and add value to an organization through the optimization of its assets and alignment with the occupiers' business. We have designed the book to bridge the gap between Strategic Management, Asset Management, Facilities Management and Contemporary Property Management, with an occupier focus. Ultimately, this book aims to raise the awareness of how real estate can support the business, transform the workplace and impact upon people and productivity.

The book is aimed at undergraduates (final year students) on General Practice Surveying, Estate Management and Facilities Management courses – plus any business courses that have real estate or facilities within them. The book is also suitable for postgraduate students undertaking a fast-track conversion course on Property, Real Estate, Asset Management, Facilities Management and Corporate Real Estate, plus MBAs – both generic and property specialist. We feel the emphasis on implementation through contemporary solutions to problems and inspiring added value action plans makes this book equally attractive to practitioners involved in tenant representation, facilities management and occupier solutions, and to occupiers such as Property or Real Estate Directors.

Deciding the content and structure of the book provided us with a considerable challenge since we realized that what we included in the book had to represent what we felt defined CREAM. The largest challenge was keeping the book restricted to a single volume. Since it was not possible to cover everything relevant on the subject in the chapters, we aimed to identify the most salient issues. We have attempted to keep the chapters readable whilst at the same time imparting as much content as possible. At the end of each chapter is a checklist for practical guidance as well as a detailed reference list which can be used for further information on a particular topic area.

We have written and organized the chapters in a logical structure. Therefore, if readers wish to read the book from start to finish they should experience a development between the chapters. However, the chapters can also be read as stand-alone chapters if the reader wishes to know about a specific topic area. Under these circumstances the reader may be directed to other parts of the book that would act as a support for that particular area.

It is our intention to provide the widest coverage possible of CREAM. We have adopted a contemporary approach in that we have crossed a number of professional boundaries. Included in this book are concepts from business management, real estate management, property management, asset management, facilities management, workplace management and environmental psychology. We believe the connections made between these different professions make this book unique.

The Layout of the Book

In Chapters 1 and 2 we aim to provide a strategic context. Chapter 1 explores issues relating to the business environment. The chapter aims to illustrate that any CREAM strategy needs to be set against the context of the changing business environment. It identifies the major trends that can impact on CREAM, with specific attention given to the multi-generational workforce. Chapter 1 demonstrates the use of traditional strategic analytical tools and more contemporary scenario planning techniques. Chapter 2 develops the strategic analysis further into the organizational context. This chapter illustrates that strategic CREAM solutions require integration between corporate strategy and the CREAM strategy. Chapter 2 demonstrates how CREAM can not only reduce costs to the organization but actually add value.

In Chapters 3 and 4 we explore procurement options. Chapter 3 explores procurement options relating to the real estate whilst Chapter 4 looks at the procurement options that relate to service delivery. Chapter 3 provides an overview of the strategic procurement options available to an organization. Included in this chapter are the issues relating to freehold, leasehold and the impacts of a sale-and-leaseback. Chapter 4 illustrates the benefits of developing a customer focused strategy.

In Chapter 5 we explore the range of different performance management and benchmarking techniques. This chapter identifies the benefits gained by benchmarking whilst at the same time illustrating the potential problems relating to comparisons.

In Chapters 6 and 7 we consider issues concerning the workplace. Chapter 6 develops a historical perspective as to the changing nature of the office work. Since work is no longer restricted to purely working in an office environment, issues relating to out of office working and virtual working are explored. The latter part of Chapter 6 focuses on productivity and the role of the behavioural environment. Chapter 7 describes a methodology for corporate relocation. Emphasis is placed on assessing organizational demand then evaluating property supply with the aim of achieving optimum alignment.

In Chapter 8 we deal with issues relating to sustainability and corporate social responsibility. The drivers of sustainability are explored and initiatives such as

BREEAM and green leases are evaluated. Specific emphasis is placed on sustainability in practice.

Terminology Used in the Book

When deciding the title of the book we considered the terms currently in use for this professional area. During our review of the literature we realized that the term Corporate Real Estate Management (CREM) was in common use. However, we felt that simply using the term 'corporate real estate management' did not cover the full range of activities that we wished to include. We felt it important to include asset management in the title as we believe that organizations need to consider their real estate as an asset rather than a liability. By shifting the emphasis to real estate as an asset then the dialogue and communication with an organization can relate to asset maximization. This approach acknowledges that real estate is an asset in a financial sense, included on the balance sheet, whilst also an asset in the operational sense and can lead to organizational performance.

Where the literature refers to Corporate Real Estate Management (CREM) we have used this term to ensure that we accurately report the work of other authors and to ensure consistency. However, where we wish to present our own views on the topic area we have adopted the term Corporate Real Estate Asset Management (CREAM).

The Limitations of the Book

The authors apologize and accept full responsibility for any errors that may have slipped through. The book is not intended to give professional advice and should not be taken as anything other than a general guide to methodology. The authors cannot accept responsibility for the results of any action taken or advice given as a result of reading the book. The addresses of Internet sites given in the text were alive at the time of writing the book but inevitably some may cease to operate. We determined a cut off period for the literature and references used in the book as November 2009.

B.P.H.
N.N.

Acknowledgements

Undertaking the writing of a textbook requires the contribution of a large number of people. We would like to thank all those who have helped and contributed to the writing of this book. We thank in particular:

Andrew Procter and Bill Fennell, Actium Consult
Rob Yates, Customer Services Director, Bruntwood
Johanna Ihalainen, Workplace Solutions Manager, Finland & Nordics, Nokia
Clive Wilkinson Architects, for permission to reproduce some of their photographs
Danny Shiem-Shin Then, for supplying the Hong Kong relocation case study
Creative Sheffield, for their collaboration in creating the objective relocation decision tool
Petr Zalsky, Europolis, Prague, for the Research at Amazon Court
Yat Tang, for work in Photoshop to create some of our images
Mandy Nunnington, for her proofreading skills during the numerous drafts created for this book
Our families for their encouragement and assistance

And finally thanks to the many colleagues and friends who provided ideas, information and knowledge to use in this book, and the hundreds of students we have taught and who have indirectly contributed to the ideas and concepts presented in this book.

Acknowledgements

Chapter 1

The Business Environment

Introduction

Constant change, and increasingly the rate of change, in the business environment means that in order for organizations to prosper, or even survive, they have to be agile and flexible enough to adapt to the changes around them. Set against this context it is understandable that corporate real estate asset management (CREAM) has to be proactive rather than reactive. Attempting to forecast future events is clearly a complex task. However, there are a range of tools and techniques at the CREAM manager's disposal which can help in providing corporate real estate asset management solutions.

In this chapter we explore the significant drivers, issues and trends in CREAM. Traditional strategic analysis tools will be presented with the emphasis on their application in corporate real estate asset management. We conclude with a strategic planning technique called scenario planning. Through a number of studies the application of scenario planning will be explored in the context of CREAM.

Drivers, Issues and Trends

In this section we explore the significant drivers, issues and trends in CREAM, facilities management and workplace management. The major significant trends are identified and subsequently developed in later chapters.

Corporate Real Estate Trends

The University of Reading has undertaken an annual survey of corporate real estate practice since 1993. Two major collaborators with the University of Reading were Johnson Controls Incorporated (JCI) and CoreNet. The annual survey was distributed by post up until 2002 and then was subsequently administered via the Internet. Each year amendments have been made to the survey questions to represent changes in corporate real estate practice. However, a number of core questions have remained unaltered.

Corporate Real Estate Asset Management. ISBN: 978-0-7282-0573-4

The collection of corporate real estate information over such a long time period enabled a detailed analysis of the major trends in corporate real estate to be undertaken (Gibson & Luck, 2006). A longitudinal analysis of the annual survey of corporate real estate practice meant that the variables could be clustered together to identify major trends in corporate real estate management in a technique called principal component analysis. This data reduction technique allowed the identification of underlying themes in corporate real estate management. Gibson & Luck (2006) identified four principal components that represent four major trends in corporate real estate. These include:

- **New working practices.** The variables that loaded onto this component revolved around desk sharing, hotelling and teleworking. This theme clearly links the changing working practices and the potential impact on corporate real estate provision. People are more dynamic when working in an office environment and also are working with increased mobility away from the office environment.
- **Outsourcing.** The debate as to whether a service should be provided in house or out of house is clearly captured in this theme. Organizations, whose prime activities are not corporate real estate, have to identify whether services should be provided within the organization or should be outsourced to service providers.
- **Technology infrastructure.** This theme captures the impact of information technology. This has two major dimensions. First, the adoption of information technology by the corporate real estate department to enable an efficient delivery of service. Secondly, the adoption of information technology by the organization, which could be closely linked to the adoption of new working practices.
- **CRE management and strategy.** The variables clustered onto this theme demonstrate a shift to a more strategic approach to corporate real estate.

The longitudinal analysis of corporate real estate practice undertaken by Gibson & Luck (2006) established four major themes (trends) in corporate real estate management. One of the major findings was that the new working practices theme was predicted to have a significant impact in future corporate real estate practice.

Facilities Management Trends

In 2007, IFMA (International Facility Management Association) undertook a piece of research to explore the current trends and future outlook for facility management professionals (IFMA, 2007). The research entailed inviting a panel of experts from the facility management profession to a two-day conference which was to be held in Houston, Texas. A total of eight invited participants took part in the forecasting session. The outcome of the forecasting session was the identification of the top trends and drivers that facility management professionals may face in the coming years. The major trends in facilities management (FM) are listed in order of importance:

- **Linking facility management to strategy.** This trend establishes an ever-clearer linkage between the FM provision and the organization's core business strategy. It includes the relationship between the physical facilities and the workplace culture and branding. It also relates the working environment to employee satisfaction and productivity.
- **Emergency preparedness.** This trend is all-encompassing and could be labelled as risk assessment. At one end of the scale are located issues such as basic safety and security, and at the other acts of terrorism and natural disasters are included.
- **Change management.** This trend relates not only to the changes that occur within the facility management department but also the changes that occur to the organization's core business. In response to changes in the facility management department this can include a continued demand to increase efficiency and responses to changing legislative requirements. The facility management department should be agile and be able to respond rapidly to changes in core business activities.
- **Sustainability.** This trend is demonstrating a significant growth in importance. On a technical level it includes energy management and evaluation of building performance. On a managerial level it includes the management of programmes that aim to reduce, reuse, recycle and are consistent with a firm's corporate social responsibility stance.
- **Emerging technology.** Technology is having an impact on the way that buildings are monitored and used. This includes complex building systems that can measure and monitor the performance of the building. The impact of technology means that people have the flexibility to work wherever they require, whether inside or even outside of a building.
- **Globalization.** The world has increasingly become a smaller place, with large organizations having a global presence. The geographical diversity of an organization requires the facility management department to respond to differences in culture, language, workplace expectations, laws and regulations.
- **Broadening diversity in the workforce.** This trend captures the changing workplace demographics. It identifies that the workplace has a multi-generational workforce, which includes an ageing workforce. The range and diversity of the workforce means that generational work styles need to be accommodated in any workspace provision.
- **Ageing buildings.** This trend captures the need for maintenance of buildings and issues relating to the replacement of the building. As a building reaches the end of its planned working life decisions have to be made by the facility management professionals as to whether to repair, reuse or replace the building.

It is interesting to note that IFMA identified the most important trend in FM as the linking of facility management to the organization's business strategy. This trend is supported by the research of Gibson & Luck (2006) in relation to corporate real estate. The alignment of corporate real estate management and FM with the corporate strategy will be specifically discussed in the next chapter.

Workplace Management Trends

There are a number of major trends impacting on the way in which people work. These drivers for change offer opportunities to rethink the way work is undertaken (Jones Lang LaSalle, 2008, p. 1).

The major workplace trends include:

- **Distributed workforce** – increasing use of technology means that work can be undertaken anywhere, even on the move.
- **Global urbanization** – the proposal that knowledge clusters are created because people tend to congregate with like-minded people.
- **Multi-generational workforce** – potentially four generations of workers could be in the same office environment. Each generation may have their own preferred work style.
- **Sustainability** – there is a trend and an expectation that the office environment will be a sustainable workplace.
- **Web. 2.0** – increasing use of online collaborative software as a means of knowledge creation and knowledge transfer.
- **Mandate for choice** – the increasing demand for flexibility means a need for a range of work settings and work styles.

The major workplace trends identified give an indication of the increasing complexity involved in designing working environments. The developments in technology mean that work is no longer restricted to the office environment. This freedom from the office means that office workers have the flexibility to work in a range of different working environments. Establishing the balance between work undertaken in an office environment and work undertaken outside the office environment is one of the challenges faced by workplace designers.

Significant Themes

Table 1.1 summarizes the major trends identified in corporate real estate, FM and workplace management. It can be seen that workplace transformation and ICT and real estate are a common theme across all the professions.

TABLE 1.1 Significant Trends in Corporate Real Estate Asset Management

	Corporate real estate	Facilities management	Workplace management
Workplace transformation	New working practices	Linking facility management to strategy	Distributed workforce mandate for choice
Strategic alignment	CRE management and strategy	Linking facility management to strategy	
ICT and real estate	Technology infrastructure	Emerging technology	Web 2.0
Globalization		Globalization	Global urbanization
Change management		Change management emergency preparedness	
Multi-generational workforce		Broadening diversity in the workforce	Multi-generational workforce
Sustainability	Consistency of real estate with corporate social responsibility Carbon reduction commitment legislation	Energy efficiency	More effective use of space
Real estate procurement	Outsourcing Changes to International accounting standards		

The major trends identified in Table 1.1 can be considered as significant themes that link together the different professional components of CREAM. The multi-generational workforce theme is related to the changing demographics of the workforce and will be developed further in the next section. Each of the other significant themes will be discussed in greater detail in subsequent chapters.

Multi-generational Workforce

The changing demographic trends mean that for the first time there is a possibility that four generations of people could be working alongside each other in today's workplace. The different generations can be categorized as four distinct groupings, each having specific workplace expectations and requirements. The four generations can be categorized as follows (Hammill, 2005):

- Born between 1922 and 1945. This generation can be classified as Veterans, Silent or Traditionalists.
- Born between 1946 and 1964. These are the Baby Boomers.
- Born between 1965 and 1980. This category is the Generation X workforce. Sometimes referred to as the Gen Xers.
- Born between 1981 and 2000. The newest entrants to the workplace come from this grouping, which is classified as Generation Y. Sometimes referred to as Gen Y, Millennials or Echo Boomers.

To ensure that workplace requirements and expectations are met there is a need to establish how each generation prefers to work for maximum productivity. In addition to creating specific workspaces for each generation it is also important to acknowledge that an office environment must allow interaction and collaboration between the different generations. Therefore, when creating a workspace for all the generations spaces should be created to enable the generations to interact efficiently, effectively and productively.

To enable the right mixture of workspaces to be created it is important to establish the different generations' attitude to work. A summary for the different work styles is presented in Table 1.2 (Hammill, 2005).

Table 1.2 clearly illustrates the range of diversity and needs of the different generations. In some of the categories there is the potential for workplace tensions. As it can be seen, Generation Y see work as a means to an end. In contrast, the Veterans' view of work is that it is an obligation. Clearly both generations have different interpretations of what work means to them. Additional tensions can be created by the means of communication in the workplace, the Veterans preferring individual work style whilst Generation Y prefer a more collaborative, participative work style.

Generation Y

Since this generation is new to the working environment, research and understanding of this generation is still being gathered. An organization that is undertaking research to establish a better understanding of the Generation Y workforce is Johnson Controls Inc. One of their global workplace innovation projects is the OXYGENZ project, which aims to establish the first global survey of Generation Y and the workplace of the future. OXYGENZ research aims to specifically establish how important the workplace is in attracting, recruiting and retaining Generation Y workers.

TABLE 1.2 Workplace Characteristics

	Veterans (1922 to 1945)	Baby Boomers (1946 to 1964)	Generation X (1965 to 1980)	Generation Y (1981 to 2000)
Work ethic and values	Work hard Respect authority Sacrifice Duty before fun Adhere to rules	Workaholics Work efficiently Crusading causes Personal fulfilment Desire quality Question authority	Eliminate the task Self reliance Warrant structure and direction Sceptical	What's next Multi-tasking Tenacity Entrepreneurial Tolerant Goal orientated
Work is ...	An obligation	An exciting adventure	A difficult challenge A contract	A means to an end Fulfilment
Leadership style	Directive Command and control	Consensual Collegial	Everyone is the same Challenge others Ask why	To be determined
Interactive style	Individual	Team player Loves meetings	Entrepreneur	Participative
Communications	Formal memo	In person	Direct Immediate	E-mail Voicemail
Feedback and rewards	No news is good news Satisfaction in a job well done	Don't anticipate it Money Title recognition	Sorry to interrupt, but how am I doing? Freedom is the best reward	Whenever I want it, at the push of a button Meaningful work
Messages that motivate	Your experience is respected	You are valued You are needed	Do it your way Forget the rules	You will work with other bright, creative people
Work and family life	Ne'er the twain shall meet	No balance Work to live	Balance	Balance

Some of the key characteristics of the Generation Y workforce have been categorized by the OXYGENZ team as follows (Holden & Pollard, 2008):

- Generation Y workers are scarce in the workplace, with not enough of them becoming part of the workforce. It is estimated that there are 1.7 billion Generation Y worldwide. This represents 26.92% of the worldwide population.
- They are transformational. The ability to be constantly connected through the Internet and mobile devices means that they are transforming both social behaviour and the way that business is undertaken.
- They do things differently. They have a multi-skilling approach using a number of digital devices at the same time. They are sometimes called 'data jockeys'.

- They are challenging. They require meaningful work and wish to be consulted with regards to management decisions. They do not like ambiguity and want clear direction and immediate feedback on their performance.

Initial findings of the project indicate that Generation Y workers wish to establish some identity with their workplace, with 85% of respondents wanting to personalize their workplace. In addition, 65% of respondents want their own desk, which is another clear signal of wishing to establish their connection to the organization. The space allocated to the Generation Y worker should not be less than 4 m^2, since only 2% of respondents feel comfortable in that amount of space. However, 81% of respondents want to work in a mobile way. This means there is clearly a balance to be struck between the number of dedicated desks and flexible mobile workers.

When the Generation Y workforce were asked about their choice of company, and creativity and productivity in the workplace they identified the following (Holden & Pollard, 2008):

- What are your three most important factors in choice of company?
  - First: Opportunities for learning
  - Second: Work colleagues
  - Third: Advancements and promotions
- What do you think makes people creative in your working environment?
  - First: People around
  - Second: Ambience and atmosphere
  - Third: Technology
- What do you think makes people productive in your working environment?
  - First: Technology
  - Second: People around
  - Third: Ambience and atmosphere

The results start to create a picture for technology-capable Generation Y workers. It is clear that they feel it is important to have the right people around them to work with, supported by the right ambience and atmosphere. The Generation Y worker is motivated by learning opportunities and progressive career enhancement when choosing a future company.

The key implications of the findings of the OXYGENZ project for corporate real estate, FM and the workplace were identified by Dr Jay L. Brand at the Dallas 2009 CoreNet Global Summit as follows (Holmes, Quick & Brand, 2009):

- **Location, location, relocation?** The majority of Generation Y workers (79%) would choose to work in another location with access to good public transport. However the majority of the Generation Y workforce (56%) still want to go to work in the car. This means there is a requirement for car parking facilities.

- **HQ and satellite offices.** The Generation Y workforce identifies three different levels of office provision. The main office building provides accommodation for permanent administrative staff and satellite offices are available for team working and team meetings. In addition there is a requirement for hired workspace, which is used on an ad hoc basis.
- **New workspace model.** Whilst the open working environment is seen as the norm, there is still a requirement to provide individual desks. In addition low density office environment should be provided, since 61% would feel comfortable in a space of 12–16 m^2. The working environment should enable collaborative working. The Generation Y workforce sees the use of technologies such as Web 2.0 as becoming the norm. Teleworking is forecast to increase.
- **A service-focused delivery.** The majority of Generation Y workforce (80%) tends to prefer modern workplace interiors with subtle, clinical and relaxing colours. There is an expectation that communal spaces will be provided on the site. In addition on-site services such as shops and retail units are expected to aid convenience.
- **Workplace.** The Generation Y workforce anticipate new ways of working to include flexible and non-traditional work styles and practices. They see the workplace as having many uses such as: working, meeting, collaborating, socializing and entertaining. The longer presence at the office means that the on-site services also need to be available.

Meeting the demands of the Generation Y worker is clearly going to be a challenge for the CREAM manager. However, the Generation Y worker is only one element of the multi-generational workforce and therefore consideration also needs to be given to the ageing workforce.

Ageing Population

As we progress into the 21st century there will be an increasing shift in the age profile of the workforces of many developed countries (Erlich & Bichard, 2008). The demographic shift to a more ageing population is driven by a number of factors. These factors can be classified as follows (Smith, 2008):

- **Shrinking pension funds.** The shortfall in the pension funds means that people can no longer retire at the normal retirement age.
- **Retained knowledge.** The older workers' knowledge and experience, which has been gathered over a lifetime, can be a valuable asset to organizations.
- **Legislation.** The ageing workforce is increasingly protected by discrimination legislation.
- **Living longer.** People are generally living longer owing to a number of factors, one of which could be developments in medical sciences.

Since forecasts indicate that the future workforce will increasingly be made up of workers over the age of 50, there is clearly a need to establish the specific workplace needs of this category of workers (Erlich & Bichard, 2008). A piece of research that specifically aims to identify the needs of the older knowledge worker and the implications for workplace design is the Welcoming Workplace project, undertaken by the Helen Hamel Centre at the Royal College of Art in the UK (Smith, 2008). The research methodology adopted by the project was case study analysis. Three case studies were investigated in different geographical locations and different industrial sectors, as follows (Smith, 2008):

- **Case study 1.** A pharmaceutical company in London, UK. The lead research partner was the Royal College of Art.
- **Case study 2.** A technology company in Yokohama, Japan. The lead research partner was Kyushu University.
- **Case study 3.** A financial services company in Melbourne, Australia. The lead research partner was the University of Melbourne.

The major findings of the Welcoming Workplace project can be summarized as follows (Smith, 2008):

- **Support for different work styles.** With the increasing trend towards collaboration through open plan work environments, there has been a loss of individual private space for concentration. In addition, spaces that allow contemplation should be considered.
- **Support for psycho-social requirements.** These findings acknowledge the changing physical and mental state of the older knowledge worker and recommend possible changes in the physical environment that could support the health and well-being of the older knowledge worker.

The ageing office knowledge worker is a product of the development of a knowledge economy and the increasing age demographic profile. As such ageing knowledge workers are predicted to become an increasing part of the future workforce, if organizations wish to attract, and retain, this category of worker then future workplace designs will have to meet their specific needs and expectations.

Multi-generational Headquarters

In April 2007 a new headquarters was opened by PricewaterhouseCoopers (PwC) in Dublin, Ireland. The 1800 workers had an age profile as shown in Table 1.3.

Table 1.3 identifies a significant number (62%) of the workforce are under the age of 30 and the remaining 38% of the workforce are above the age of 30. Whilst the age profile reflected predominantly younger workers, the age range is significant enough

TABLE 1.3 Age Profile of PwC Dublin Workforce

Age range	Percentage of total
Under age 30	62
In their 30s	24
Over 40	14

for the need to consider multi-generational working. The issues that this raised for PwC were identified as (Hughes & Simoneaux, 2008):

- How do you effectively manage a workforce of diverse ages and expectations?
- How do you plan and build a workplace that performs for all ages?
- What are the best ways to facilitate the transfer of huge stores of accumulated business knowledge from older to younger workers?

The transfer of knowledge from older to younger workers is an essential component of any knowledge creation and knowledge transfer organization. One way that younger colleagues can learn is by overhearing their older colleagues' conversations. PwC identified the need for this transfer of knowledge from older to younger workers, but acknowledged the potential tension between potentially older workers preferring individual private offices and younger workers requiring open plan office environments (Hughes & Simoneaux, 2008).

The creation of a new headquarters was also an opportune time to consider issues relating to how PwC wanted to define itself as a business. This meant clarifying what PwC wanted to create in the new workplace with regards to new organizational culture and new business practices.

One issue identified by the younger workers related to the lack of transportation from their new headquarters location, Spencer Dock, to Dublin's dynamic urban culture and entertainment scene. Since PwC's wanted to attract and retain young talented people they provided their own buses to supplement existing public transport.

The open plan workspaces were created with input from people from every level of the organization. Mary Cullen, a member of the PwC senior partner team, commented: 'It got people to think about what sort of space they needed to do their work' (Hughes & Simoneaux, 2008, p. 35). This led to a range of different workspaces being provided which enabled flexible and different workstyles to be adopted. This approach links to the proposals by the Welcoming Workplace project to provide workspace to collaborate, concentrate and contemplate (Smith, 2008).

The PwC headquarters attempts to balance formal and informal interactions:

> *People meet and talk in both formal and informal spaces throughout the building. They also enjoy a 200-seat restaurant, Starbucks coffee dock, fitness*

centre, and state-of-the-art training and meeting rooms. Walkways and bridges provide easy access to every part of the three-block-sized building.

Hughes & Simoneaux, 2008, p. 35

Keeping everyone informed in the developments of the new headquarters was seen as key to the success of its implementation. This led to the creation of a communications strategy which started two years before the opening of the building. Representatives from every business unit were included in the communications strategy with the intention of information being communicated both ways. The communication strategy included a number of different methods, including (Hughes & Simoneaux, 2008, p. 36):

- Online surveys of all PwC people on issues such as transportation, the restaurant, and fitness centre.
- An intranet site with regularly updated information and a Q&A forum.
- Regular meetings with a focus group of representatives from each department, who in turn circulated information via email and in person to everyone in their groups.
- Individual and team interviews at all levels of the organization.
- Short video presentations by Donal O'Connor, the organization's chief executive, so he could speak with everybody.
- A printed reference guide that replicated the intranet site, published shortly before the move to the new building.
- A show area set up on site nine months before the move.
- 'Showcases' with all suppliers for the project, from furniture and technology to the new cashless vending system, two months before the move.

There are a number of lessons learned from this multi-generational case. First, it is important to start by understanding the principal aim of the business and organizational culture that you wish to achieve. Secondly, detailed research and investigation is required to establish, and understand, different occupier requirements. Thirdly, new workplace designs should incorporate diversity to enable a range of different working practices. And finally, communication should be started as soon as possible and remain constant throughout the project. This enables office occupiers to contribute to the designs and be informed of any new developments. (See the Nokia case study, p. 253, for further illustration of the linkage between communication and workplace management.)

Strategic Analysis Tools

In this section we will introduce a number of strategic analytical techniques that can be used by the CREAM manager to establish and evaluate the potential impacts of the main drivers for change in the business environment. The tools and techniques offer the CREAM manager a systematic framework for evaluation of the business environment. The strategic analytical techniques include: PESTEL analysis, SWOT analysis and competitor analysis. These techniques are included to broaden the understanding of what environmental influences there are and how these affect real estate and its management.

PESTEL Analysis

When considering the future strategy of an organization, or the CREAM department, it is useful to start by identifying external factors that may have an impact on that future strategy. The PESTEL analysis offers a structured way of identifying the following external factors:

- **Political** – includes the impact of government policy and regulation. Changes in planning policy may have a direct impact on future real estate strategies.
- **Economic** – relates to changes in interest rates, exchange rates and availability of money supplies, an important consideration when considering property procurement.
- **Social** – embraces the changing workplace demographic, which includes population growth. Also included are issues relating to cultural differences and an ageing population, which need to be considered during workplace design.
- **Technological** – includes factors that enable organizations and employees to use technologies to enable them to be more productive and profitable.
- **Environmental** – embraces factors that relate to energy efficiency, recycling and sustainability. Links directly to sustainable developments and green buildings.
- **Legislative** – the factors that create the legal boundaries in which organizations have to work.

Variations on the PESTEL analysis include: PEST analysis (Political, Economic, Social and Technological) or STEP analysis (Sociological, Technological, Economical and Political.

STEP Analysis: Illustrated Example

Dr Nigel Oseland used a STEP analysis, see Figure 1.1, to evaluate the changes in work style and work environment to meet changes in global markets and economies (Oseland, 2008).

The STEP analysis identified the following external factors:

- **Social.** Oseland identifies how the changing demographic leads to potential tensions in the multi-generational workplace. New entrants into the workplace have different

(Continued)

Political	Economic
Sustainability and CSR European legislation Security and terrorism Mixed use	Market demand Location/grouping Overseas outsourcing Business timescales
Social	**Technology**
Socio-demographics Longevity and retirement age Attitude to work Work–life balance	Flat screen Wireless and mobile devices Broadband and home-working Thin client

Figure 1.1 Work style and workplace STEP analysis. *(From Oseland, 2008; reproduced with permission)*

expectations than the 'Baby Boomer' generation. Ensuring equality within a workplace context means that workplace designs have to include provision for disability, culture and ethnicity.

- **Technological.** Included in this factor is the range of technological tools available in today's working environment. Included are laptops and PDAs with wireless connectivity which gives people flexibility and mobility. The impact of the Internet means the development of social networking sites, which leads to a change in how people relate and interact with each other. Oseland identifies that the physical working environment needs to be interesting and exciting if it is to attract people back into the office environment.
- **Economic.** In response to changing global markets and economies the workplace environment has to be flexible and agile, enabling it to respond quickly to changing business requirements. Global virtual working enables teams to collaborate, thereby reducing time to market.
- **Political.** Oseland includes in this factor issues relating to sustainability and social corporate responsibility. With most offices utilized for only 50% of the time then significant savings in embodied energy can be made by reducing the building space provided. In addition, good design and management can further reduce energy consumption and the impact on the environment.

On completion of the STEP analysis Oseland concludes that work is becoming a balance between the physical and virtual world, requiring seamless interconnection. The office worker is becoming more international and multicultural, working to performance output rather than time spent in the office (Oseland, 2008).

SWOT Analysis

The SWOT analysis is a very simple but powerful analytical tool. It enables a strategic review by evaluating both the internal and external factors affecting an organization. The internal factors are identified by the strengths (S) and weaknesses (W) of an organization. The external factors identify the opportunities (O) and threats (T) facing the organization. In essence a SWOT analysis enables a company to evaluate how well it is positioned, through its strengths and weaknesses, to respond to the external factors which are the opportunities and threats.

It should not be seen as a mere identification and listing process but as a structured analysis that brings together the strategic capabilities of an organization against the business environment and when used properly provides a powerful discussion framework which can highlight matches between the strategic capability of an organization and opportunities that could be exploited. It can also highlight the exposure of weaknesses in the company to major threats in the environment.

The SWOT analysis can be undertaken at both macro and micro levels. The macro level SWOT enables a corporate strategic overview, whilst the micro level SWOT may be appropriate for an individual business unit or property.

An example of a simple descriptive SWOT for an individual property within a corporate portfolio review, in this case for a food manufacturing company, is shown in Figure 1.2.

It is clear that this building, which was acquired organically through expansion may no longer support the core business. The SWOT should lead to consideration of

Strengths	**Weaknesses**
Freehold property acquired as part of acquisition of a business No mortgage or other outstanding debts	High occupancy costs to operating costs ratio Low space utilization High occupancy costs per person Difficult access to site, not designed for latest articulated vehicles Location is long distance from processing plant increasing food miles and reducing sustainability Planning restrictions due to proximity to residential use restricts working hours
Opportunities	**Threats**
Potential alternative use value for residential use Unique site in sought-after area Significant uplift in value if planning permission obtained for residential use Expectation of house builders returning to acquire land in next few years	Old buildings are maintenance liability and may not comply with ever more stringent health and safety legislation Poor energy performance and incompatible with company sustainability policy Current weak market for residential land

Figure 1.2 An illustrative SWOT for a single asset within an occupier's portfolio

strategic options to manage the threats to the business and capture the opportunities for disposal for an alternative use.

A variation on the SWOT analysis is the Scored SWOT analysis (Johnson & Scholes, 1999). The Scored SWOT analysis allows the strengths and weaknesses of the organization to be rated against the factors in the external business environment.

How to undertake a Scored SWOT analysis:

- Identify the current strategies the organization is following.
- Identify the key changes anticipated in the environment.
- Identify the key strengths and weaknesses of the organization.
- Produce a matrix of strategies, strengths and weaknesses against the key environmental issues.
- Examine each strategy, strength and weakness against the key environmental issue.
- Score the outcome:

+ +	a SIGNIFICANT benefit to the company
+	a benefit to the company
0	no impact on the company
–	an adverse effect on the company
– –	a SIGNIFICANT adverse effect on the company

- Total the scores to identify the linkages between the issues in the environment and the strengths and weaknesses of the organization/sector.

Figure 1.3 demonstrates the Scored SWOT approach for a regional out of town shopping centre.

Competitor Analysis

The CREAM service provision can be provided by in-house staff or by an outsourced service provider. It is therefore important to constantly evaluate the CREAM service provision with potential competitors, as a way of ensuring that the core business is achieving best value service provision.

A useful tool to identify the forces affecting the level of competition in an industry is Michael Porter's Five Forces Analysis (Porter, 1980). The five forces are illustrated in Figure 1.4.

		Issues in the environment (drivers)					
	Planning restrictions on centre expansion	New major city centre scheme planned	Increased Internet based shopping.	Sustainability issues including proposals to force charging for car parking	Ageing population, the rise of the 'grey' pound	Credit crunch and potential for rising interest rates, plus increased VAT rates	
STRENGTHS							
Customer loyalty	0	+ +	–	–	0	+	+1
Resilient sales volumes	0	+ +	–	–	0	+	+1
Customer experience	–	+	0	–	0	0	–1
Pro-active adjustment of tenant mix	0	+	0	0	+	+	+3
WEAKNESSES							
Saturation of market share	– –	– –	– –	0	–	–	–8
Website	0	0	– –	0	0	– –	–4
Limited expansion opportunities	– –	– –	0	0	– –	0	–6
Youths in groups deter older customers	0	0	0	0	– –	0	–2
	–5	+2	–6	–3	–4	0	

Figure 1.3 An illustrative Scored SWOT for a regional shopping centre

(1) Potential Entrants

The threat of competitors entering an industry will depend upon the barriers to entry which include:

- economies of scale
- capital cost of entry
- access to distribution channels
- established experience and networks
- expected retaliation
- legislation/government policy/action
- differentiation and the ability to offer real or perceived added value

The diverse nature of the CREAM sector means that providers come from a range of different service backgrounds. Typical disciplines include facilities management, property management, asset management and building services.

(2) The Power of Suppliers

The bargaining power of suppliers will depend on the concentration of suppliers. If the CREAM service provision is totally provided in-house then the supply chain is

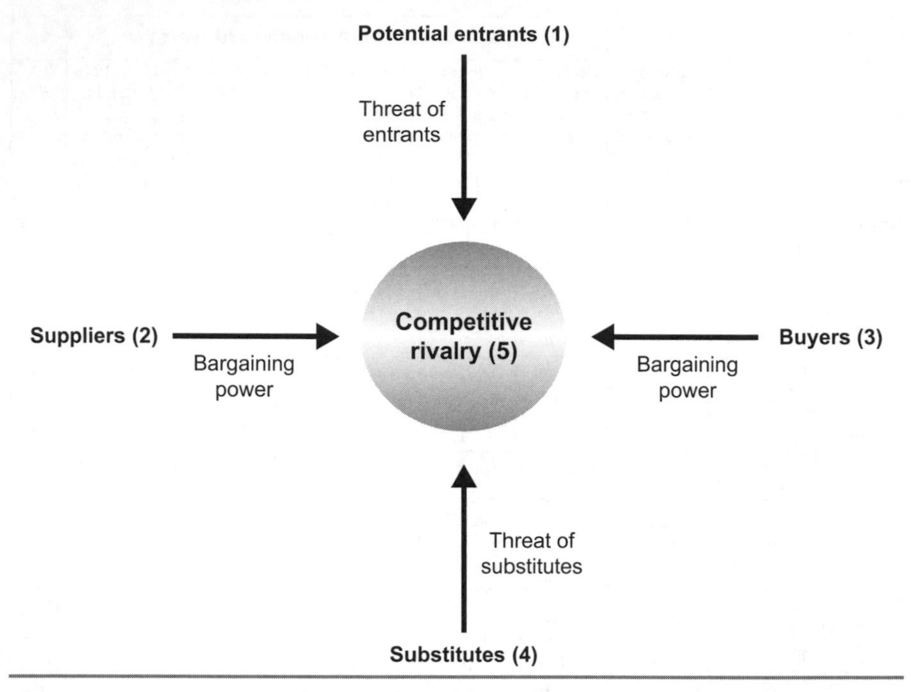

Figure 1.4 Five Forces Analysis. *(Adapted from Porter, 1980, with permission)*

shortened and the power of the supplier is reduced. However, if the CREAM services are totally outsourced then the power of the supplier is increased. Whilst the 'one-stop' shop approach for CREAM service provision means that all the services are provided by one supplier, they can demonstrate their power by ensuring that any costs of switching to another provider are high.

(3) The Power of Buyers

In a CREAM service provision context the buyer is the organization to which the CREAM services are provided. The way the power of the buyer is exerted on the CREAM provision depends on the objectives of the organization using CREAM services. If the organization is expanding then the demand for CREAM will grow. However, if the organization is contracting then the pressure to cut CREAM services with be increased.

(4) The Threat of Substitutes

The competitive position of a company can be totally undermined if the services are provided in a different way through substitution. The increasing transparency of data means that no longer is the knowledge contained within the highly paid consultant's head. Technological advancements have meant the information is more readily available. The new models of delivering CREAM services include multidisciplinary teams. The convergence in the industry means that accountancy firms and management consultancy firms are entering the market as CREAM service providers.

(5) Competitive Rivalry

These represent the existing firms competing to provide CREAM services. The nature of the competing firms will determine the threat. If the CREAM providers are of equal size then there is the potential for intense competition. The geographical coverage will have an input to the intensity of threat. If the organization requiring CREAM services is a multinational then the competition with other CREAM providers could be on a global scale.

The use of techniques such as PESTEL analysis, SWOT analysis and competitor analysis give the CREAM manager a structured approach to evaluating drivers for change. The application of such tools at departmental level ensures that the CREAM department constantly monitors and makes necessary adjustments to ensure that best value service provision is provided to the organization.

Future Scenarios

A potential limitation of the traditional strategic analysis and environmental scanning tools is that they are based on analysing the current business environment. Future planning cannot always be simply extrapolated from past and current circumstances. Planning for the future is a little more complex owing to the amount of unknown future variables and events. A strategic planning tool known as scenario planning or scenario thinking is a method that allows alternative futures to be forecast.

The scenarios generated are usually forecasts for up to the next 30 years. Each alternative future or scenario created is usually supported by a narrative or a story about that possible future. The narrative is important as it enables a deeper understanding of what that possible future would really be like.

Having identified alternative scenarios an evaluation can be undertaken to establish the potential impact each future scenario would have on an organization.

The scenario planning process enables sharing and learning between the people involved in the creation of the scenarios. In addition the scenarios created may act as a catalyst for change. The change required may be a wake-up call, which means an organization needs to undertake a step change or a paradigm shift if it wishes to thrive and prosper in the future.

The benefits of using scenario planning to forecast real estate markets are identified as (Saurin, Ratcliffe & Puybaraud, 2008, pp. 248-249):

- The future of work and pervasive influence of information and communication technologies will impact on the current real estate supply, making a lot of the current stock obsolete.
- The threat of climate change will affect locational decisions and threaten real estate portfolios but will also demand that buildings are more environmentally friendly.
- It can improve the contribution to sustainability of the construction and property investment, development and management processes, recognizing that the built infrastructure represents a substantial and relatively stable environmental resource and societal asset.
- The potential importance of corporate social responsibility to ensure higher levels of social cohesion is maintained for the longer term interests of all parties.
- The next phase of the real estate agenda is likely to focus much more attention on the relationships between employers and their employees, workers and their places of work, and occupiers and suppliers, to shape the future workplace.

The creation of possible future scenarios may be considered as a creative process. However, it is important to have some sequence or methodology to assist in the shared learning process. The Futures Academy at the Dublin Institute of Technology (DIT) has devised a scenario planning methodology, as can be seen in Figure 1.5.

The 'Perspectives Through Scenarios' methodology (Ratcliffe & Sirr, 2003) has a number of key stages. An overview of these stages will now be identified.

The Strategic Question

Identifying the strategic question is the start point for the strategic planning process. It is also the start point on the shared learning and understanding of the issues to be addressed. This is usually achieved through workshops or interviews with leaders of a particular field. The process entails strategic conversations as a way of clarifying and specifying the strategic question. The aim is to establish the vital issues that need to be incorporated into the strategic question. Implicit within the strategic question should be the identification of the expected outcome of the strategic scenario planning process.

Examples of relevant strategic questions are:

- 'What are the major forces of change affecting the global real estate industry, and how should the property profession position itself now to face the future?' (King Sturge, 2001, p. 4)

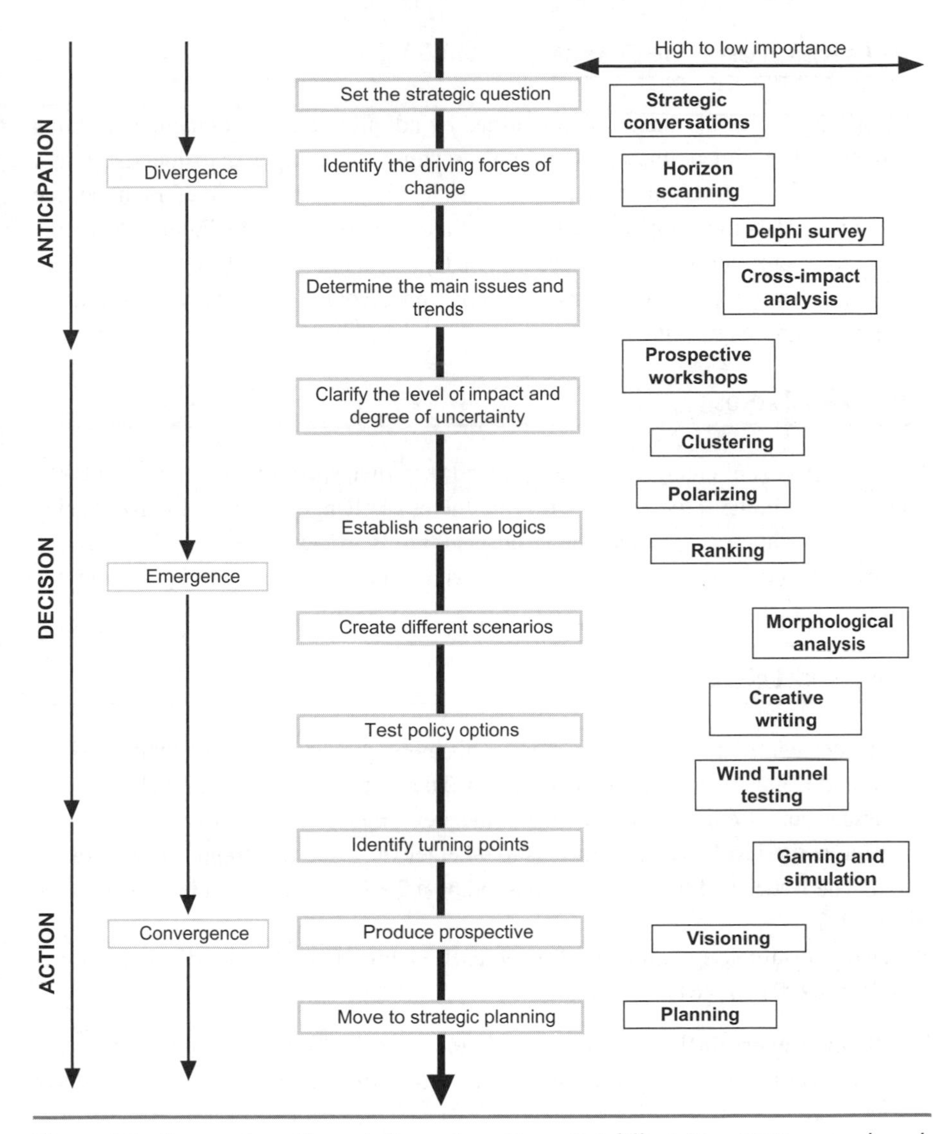

Figure 1.5 Perspectives Through Scenarios. *(From Ratcliffe & Sirr, 2003; reproduced with permission)*

- 'What are the major forces of change affecting the European real estate industry and how should the property community prepare itself now to face a future of uncertainty and complexity?' (King Sturge, 2005, p. 2)
- 'What might the future of the sustainable workplace be like in 2030?' (Saurin et al., 2008, p. 251)

The Driving Forces of Change

The driving forces of change are identified by adopting an environmental scanning technique. The aim of the environmental scanning technique is to identify major trends, issues or problems that may be in the pipeline, and sources of hope and inspiration. The environmental scanning technique adopted by the Futures Academy is the 'six sector approach'. The principle adopted is that each of the driving forces identified can be categorized into: demography, society, economy, governance, technology and environment.

Issues and Trends

Once the drivers of change have been categorized then specific issues and trends can be identified. In the same way that driving forces of change can be categorized using the six sector approach, so can the issues and trends. A clear linkage exists between the driving forces of change and the issues and trends, as the latter follows from the former.

Impact and Uncertainty

The next stage of the strategic scenario planning process is the evaluation of the issues and trends in relation to two criteria. The first criterion is the level of impact the issues and trends would have on the item under investigation. The second criterion is the level of uncertainty as to whether the issue and trend would actually occur. The issues and trends can be plotted on a 2×2 matrix, as can be illustrated in Figure 1.6.

Using terminology coined by IDON, each quadrant can be identified as follows (Ratcliffe, 2001, p. 461):

- **Pivotal uncertainties** – these are likely to have a direct impact, but their outcome is uncertain. They are pivotal in the sense that the way they turn out may have

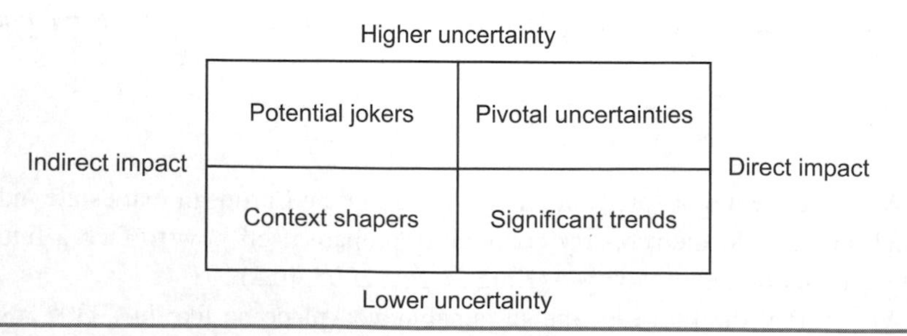

Figure 1.6 Positioning issues and trends

strong directional consequences. These are the areas that will determine the shape of different scenarios.

- **Potential jokers** – these are pretty uncertain as to their outcome and less relevant. However, it could be dangerous to treat them as mere 'noise'. They represent factors to monitor on the 'corporate radar' in case they move strongly to the right.
- **Significant trends** – these impact more directly upon the question in hand and it should be possible to anticipate their effect.
- **Context shapers** – these are relatively certain and, therefore, will surely shape the future context.

Creating the Scenarios

The previous stages collectively allow the creation of a number of alternative futures. It is at this stage that a creative element is introduced to create alternative scenarios. The scenarios created should be uniquely different thereby creating the widest diversity of possible future events.

Two pieces of research undertaken by King Sturge applied scenario planning techniques to the real estate industry. The first report attempted to evaluate the major forces affecting the global real estate industry, and the second report evaluated the major forces affecting the European real estate industry (King Sturge, 2001, 2005).

King Sturge's first report included the strategic question:

> *What are the major forces of change affecting the global real estate industry, and how should the property profession position itself now to face the future.*
>
> **King Sturge, 2001, p. 4**

Three scenarios were created in response to the strategic question.

1. **Lords of misrule:** Social reaction to over-rapid change (Ratcliffe, 2001, p. 461)
   Socio-political backlash against the forces of change.
   Regressive developments in institutions, failure of cohesion amongst the wealthy world and a dislocation in the developing nations.
   A world that moves towards increasing instability.

This scenario identifies a breaking down of social and political structures. This potentially leads to governments acting on an individual rather than a collective basis. Ultimately, the uncertainty in this scenario means that the world becomes an unsettled place.

2. **Bazaar:** Complexity managed by 'marketizing' decision processes (Ratcliffe, 2001, p. 462)
   New technologies rapidly change the fundamental principles by which industry and commerce are structured.

Social and governmental institutions are weakened at the national and international level.
Industrial nations are fragmented into many differentiated and competing sub-national regions and interests.

In this scenario the developments in technology have fundamentally changed business practices. Alliances are sought by like-minded businesses leading to a clustering of interest groups.

3. **Socratic systems:** Harnessing the knowledge economy (Ratcliffe, 2001, p. 463)
   Commercial and institutional renewal accelerates across the developed world.
   Policy-making and decision-taking become increasingly delegated and expert.
   Institutional improvements worldwide facilitate sustainable development.

This possible future is more optimistic. It presents a future where the people with expertise and the knowledge are able to make a valuable contribution. Included in this scenario is the possibility of ensuring that future developments are sustainable.

In King Sturge's second report the emphasis was shifted from a global perspective to a European perspective. It included the strategic question:

> *What are the major forces of change affecting the European real estate industry and how should the property community prepare itself now to face a future of uncertainty and complexity?*
>
> King Sturge, 2005, p. 2

The application of scenario planning methodology led to the creation of four different possible scenarios. The four scenarios created can be seen in Table 1.4 (King Sturge, 2005).

The four different scenarios provide wide-ranging considerations of possible future events. Included in the scenarios are considerations for political stability through to political instability, from economic growth to economic stagnation, from social development to social unrest.

A piece of research that applies the 'Perspectives Through Scenarios' methodology to understanding future workplace needs included the strategic question:

> *What might the future of the sustainable workplace be like in 2030?*
>
> Saurin et al., 2008, p. 251

The following three scenarios were created (Saurin et al. 2008, p. 255):

- **Jazz** – a Global Market by 2030: the workplace is a network. This scenario assumes an unprecedented acceleration of economic growth, relentless pressure for short-term gains and fierce competition on a global scale, driven by rapid technological advances and further market integration.
- **Wise Counsels** – a Secure World by 2030: the workplace is a community. This scenario assumes global economic stability and an effort to attain environmental

TABLE 1.4 Future Scenarios that May Affect the European Real Estate Industry

Empyrean: Fear and trepidation; the rise of the super-state	This scenario assumes a United States of Europe is complete but controlled by technology. Big Brother is alive and well and dominant. Age fascism is rife. Unrest among citizens is widespread and public distrust of political functions is mounting (King Sturge, 2005, p. 17)
Principia Ethica: The moral imperative	This scenario assumes a period of global metamorphosis. Further integration is a success in geographical terms as well as economic and political. Europe enjoys unparalleled economic growth through legal certainty and market transparency. A sustainable and high quality of life is enjoyed by all; the European dream has arrived and corruption has been eliminated (King Sturge, 2005, p. 21)
Titans of Avarice: Market forces on the march	This scenario assumes steady economic growth, the further opening-up of markets and rapid technological advances. EU enlargement is primarily based on economic integration, at the expense of further political unification. The green agenda has been shelved, leading to accelerated climate change and social unrest. Wealth disparities between rich and poor have increased, sparking a socio-political backlash against the forces of change. Migration and crime are at an all-time high (King Sturge, 2005, p. 24)
Belshazzar's Feast: Federal fragmentation	This scenario assumes stagnant economic growth in many parts of Europe, brought on by worldwide instability. Further integration is abandoned, while protectionist policies dominate the political agenda. The welfare state barely remains intact and has become a burden on core European countries. The European Dream is in tatters having failed to develop a common security and foreign policy. A number of regions are retreating into corrupt fiefdoms. Increasing concern arises over the threat of social unrest, conflict and environmental degradation. Crime, corruption and chaos are rife (King Sturge, 2005, p. 27)

balance and social progress. Institutional improvements worldwide facilitate sustainable development. It is a world where collective, collaborative and consensual action is favoured.

- **Dantesque** – a Fragmented World by 2030: the workplace is a fortress. This scenario assumes global economic stagnation, cultural difference and insecurity. Emphasis on distrust, retrenchment and reaction leads to widespread social unrest, conflict and environmental degradation.

Each of the scenarios has direct implications for the future of the sustainable workplace. The implications of the scenarios led to the creation of three possible future workplaces. These include the Hive, the Eco-Office, and Gattaca (Ratcliffe & Saurin, 2008).

- **The Hive** – *agility, anonymity and access*: The jazz global workplace scenario with its increased networked technology and constant drive for economic growth could mean that the corporate office becomes a thing of the past. People can work wherever they are connected.
- **The Eco-Office** – *a radical form of industrial democracy and corporate re-engineering:* The Wise Counsels' workplace scenario creates a future including global economic stability which can form the basis for considerations relating to sustainable developments. Employees can use technology to work outside of the office environment. The office environments created are of a high quality and consideration is given to issues relating to sustainability.
- **Gattaca** – *the rise of the corporate office:* The Dantesque workplace scenario incorporates the concept of the office environment as a manufacturing plant. This leads to a mechanistic view of the office and leads to the growth of the corporate office environment.

Dom Sherry, VP Strategy and Innovation at Johnson Controls, identified that when considering the future workplace some aspects of the scenarios are more probable. The most probable issues relating to the future workplace are as follows (Ratcliffe & Saurin, 2008, p. 3):

- Individuals will be increasingly networked, connected virtually and more loosely connected to corporate organizations
- Formal working environments will need to be highly collaborative to facilitate the virtual networks
- Workplaces must be:
  - highly technology enabled
  - environmentally friendly
  - offers a sense of community
  - balance physical security with a feeling of openness
- Corporates will be socially and environmentally responsible

In this section we have explored the creation of possible future scenarios for real estate and workplace. Once the scenarios have been created the next stage of the process is to test them.

Testing Policy Options

Scenarios need to be tested for robustness against the alternative possible futures. The lessons learned from the policy options test stage can be backward-engineered into current policy and strategy decision-making so that the most robust strategies and policies can be created.

Summary of Scenario Planning

Whilst the scenarios created for the questions relating to global real estate, European real estate and the sustainable workplaces do not provide specific answers, they do provide a range of possible answers. It is often claimed that it is the process of strategic scenario planning that is the most important element rather than the specific end products. The strategic planning process provides a vehicle for communication, learning and debate. It enables people to lift their heads above a normally short time horizon and take a look at the bigger picture. The strategic scenario planning methodology enables people to be creative and think the unthinkable by imagineering alternative futures.

Think the Unthinkable

To conclude this chapter we encourage the idea of thinking the unthinkable, in essence what scenario planning is designed to do. Bill Gates (2000) suggests:

> *We always overestimate the change that will occur in the next two years and underestimate the change that will occur in the next ten. Don't let yourself be lulled into inaction.*

We endorse this and use two personal examples to illustrate how hard it is to predict long term change.

Unthinkable Change

Example 1

The authors have been involved in teaching this subject in Poland for many years. When the courses started in 1995 the only flights available were with British Airways from Heathrow to Warsaw for around £280 return (at 1995 prices). In 2009 the authors made the same journey for less than £60 return from their regional airport (Doncaster–Sheffield). Not only are flights to Warsaw available from a variety of regional airports at low cost but flights are available to many secondary and tertiary cities in Poland, including Katowice, Poznan and Lodz. This illustrates not only a paradigm shift in the airline industry but other impacts of globalization such as labour migration.

Example 2

One of the authors completed his undergraduate dissertation in 1982, examining the impact of technology on real estate development appraisals. The dissertation was manually typed as word processing was not yet in the mainstream and he used a Sinclair ZX 'computer' with just one kilobyte of memory. The author had to buy additional one kilobyte chips of RAM to run a Monte Carlo risk simulation program. The alternative would have been to use punched cards on the Sheffield City Polytechnic mainframe. The

Internet was not mainstream and the very first personal computers were entering the market at very high cost.

In 2009, 25 years later, he recently spoke to his co-author in Sheffield on an Apple i-phone with eight megabytes of memory. The call was made using Skype, a voice over Internet application, via the Internet, from a Starbucks coffee shop in Istanbul. The call cost just €0.35 for 25 minutes of conversation. What is remarkable is the seamless nature of contemporary mobile technology. The i-phone logged onto the Starbucks free wireless network and the Skype application is activated by one touch of the screen. In 25 years a mobile device is capable of doing much more than a mainframe, has 8,000,000,000 times the memory capacity than the ZX81 and enables knowledge workers to work anywhere, anytime at very low cost.

As these examples show, it is difficult, but essential to think how the next 25 years may create the unthinkable.

Summary Checklist

1. Undertake a PESTEL analysis for your organization to establish key drivers affecting your business. Ensure that representatives from the strategic management team and the real estate team are present.
2. Convert the drivers into the 10 most significant key issues in the environment for your business.
3. Undertake a SWOT analysis for all your real estate assets.
4. Load the strengths and weaknesses onto a grid and compare them to the issues in the environment. This should reveal the most significant opportunities and threats and inform the CREAM strategy.
5. Consider scenario planning techniques where your business environment is especially volatile.
6. In your scenarios encourage the participants to think the unthinkable to avoid the myopia suggested by Bill Gates.
7. Review your real estate from the perspective of:
   - **a.** Its ability to support 'new ways of working'
   - **b.** Its ability to attract and retain the best staff – especially Generation Y graduates
   - **c.** The continuing impact of technology which is promoting an 'ever-connected' environment and ultimate flexibility to work wherever you want.

References

Erlich, A., & Bichard, J.-A. (2008). The Welcoming Workplace: designing for ageing knowledge workers. *Journal of Corporate Real Estate, 10*(4), 273–285.

Gates, B. (2000). *Business @ the speed of thought: Succeeding in the digital economy.* New York, NY: Time Warner.

Gibson, V., & Luck, R. (2006). Longitudinal analysis of corporate real estate practice. *Facilities, 10*(3/4), 74–89.

Hammill, G. (2005). *Mixing and Managing Four Generations of Employees.* MDUMagazine Online (Winter/Spring). <http://www.fdu.edu/newspubs/magazine/05ws/generations.htm>. Accessed 30.06.09.

Holden, G., & Pollard, S. (2008). *OXYGENZ: Envisioning the Gen Y workplace. Berlin Summit (September).* Retrieved from Corenet Global: http://www2.corenetglobal.org/dotCMS/kcoAsset?assetInode=4261335. Accessed 29 June 2009.

Holmes, M., Quick, S. E., & Brand, J. L. (2009). *OXYGENZ - Envisioning the Gen Y workplace. Dallas Summit (April).* Retrieved from Corenet Global: http://www2.corenetglobal.org/dotCMS/kcoAsset?assetInode=5713103. Accessed 30 June 2009.

Hughes, J. E., & Simoneaux, B. (2008). Multi-generational work force design: Pricewaterhouse Coopers opens a new headquaters in Ireland. *The Leader, May/June* 32–36.

IFMA. (2007). *Facility management forecast 2007: Exploring the current trends and future outlook for facility management professionals.* Retrieved from International Facility Mangement Association (IFMA). http://www.ifma.org/tools/research/forecast_rpts/2007.pdf. Accessed 28.01.10.

Johnson, G., & Scholes, K. (1999). *Exploring corporate strategy text and cases, Fifth Edition.* Harlow: Prentice–Hall.

Jones Lang LaSalle. (2008). *Perspectives on workplace*. London: Jones Lang LaSalle.

King Sturge. (2005). *European real estate scenarios: Nirvana or nemesis?* London: King Sturge.

King Sturge. (2001). *Global real estate scenarios.* London: King Sturge.

Oseland, N. (2008). The evolving workplace. *Property and Facilities Management* (October) 14–16.

Porter, M. (1980). *Competitive strategy.* New York, NY: The Free Press.

Ratcliffe, J. (2001). Imagineering global real estate: a property foresight exercise. *Foresight, 3*(5), 453–465.

Ratcliffe, J., & Saurin, R. (2007). *Workplace futures: A prospective through scenarios.* Aldershot: Johnson Controls Facilities Innovation Programme.

Ratcliffe, J., & Saurin, R. (2008). *Towards tomorrow's sustainable workplace: Imagineering a sustainable workplace future*. Aldershot: Johnson Controls Global Workplace Innovation.

Ratcliffe, J., & Sirr, L. (2003). *The prospective process through scenario thinking for the built and human environment: A tool for exploring urban futures.* Dublin: The Futures Academy.

Saurin, R., Ratcliffe, J., & Puybaraud, M. (2008). Tomorrow's workplace: a futures approach using prospective through scenarios. *Journal of Corporate Real Estate,* 243–261.

Smith, J. (2008). *Welcoming Workplace: Designing office space for an ageing workforce in the 21st century knowledge economy.* London: Helen Hamlyn Centre, Royal College of Art.

Chapter 2

Strategic Alignment

Introduction

In providing corporate real estate asset management (CREAM) strategic solutions the CREAM manager has to provide a solution that supports the organizational requirements. This requires a business focus with linkages between the corporate strategy and the CREAM strategy being fully integrated. Providing the linkages and establishing how the CREAM provision can add value to the organization requires the CREAM manger to be fully conversant with the principles and application of the development of corporate strategy.

In this chapter we will demonstrate that a strategic CREAM solution requires the integration of a number of professional areas. The role of corporate strategy will be explored with specific emphasis given to how CREAM can add value to the organization. We will conclude with a model for CREAM strategic alignment.

An Integrated Approach: Collapsing of Boundaries

To understand the processes and practices of CREAM it is important to have clear definitions. Unfortunately, as the disciplines that are part of CREAM have grown the boundaries between them have become blurred. This section attempts to clarify the definitions and track the changing nature of CREAM and the expectations of organizations.

Asset Management

In 2008, the Royal Institute of Chartered Surveyors (RICS) produced guidelines for asset management for the public sector. Included in these guidelines are two useful definitions relating to asset management. The first relates to what constitutes the asset base, which is defined as:

> *The entirety of the land and building assets owned or occupied by an organization.*

Jones & White, 2008, p. ix

Corporate Real Estate Asset Management. ISBN: 978-0-7282-0573-4

The second definition relates to asset management and is defined as:

The activity that ensures that the land and buildings asset base of an organization is optimally structured in the best corporate interest of the organization concerned. It seeks to align the asset base with the organization's corporate goals and objectives. It requires business skills as well as property skills although only an overall knowledge of property matters is required. However, property input within the overall process is imperative.

Jones & White, 2008, p. ix

This definition of asset management clearly identifies the need to link the land and buildings of an organization to the core business objectives. It also makes clear that professionals working in this area need to have both business skills and an overall knowledge of property matters.

Property Management

Property management is a term that tends to be used in the UK and Europe, whereas real estate management is used by the rest of the world.

Property management can be considered to be the day-to-day operational aspects of the organization's properties.

It is sometimes referred to as 'operational' and it is the activity of undertaking the professional/technical work necessary to ensure that property is in the condition desired, in the form and layout and location desired and supplied with the services required, together with related activities such as the disposal of surplus property, the construction or acquisition of new property, the valuation of property, dealing with landlord and tenant and rating matters, all at an optimum and affordable cost.

Jones & White, 2008, p. ix

Facilities Management (FM)

Facilities management is the term used in the UK and Europe, whereas facility management tends to be used by the rest of the world.

The British Institute of Facilities Management (BIFM) has formally adopted the definition of FM provided by CEN, the European Committee for Standardization, and ratified by BSI British Standards:

Facilities management is the integration of processes within an organization to maintain and develop the agreed services which support and improve the effectiveness of its primary activities.

BIFM, 2009

This definition as it stands appears a little vague, since the terms 'processes' and 'agreed services' are not expanded upon. However, the BIFM adds the following:

> *Facilities management encompasses multi-disciplinary activities within the built environment and the management of their impact upon people and the workplace.*
>
> BIFM, 2009

This additional definition of FM starts to establish that facilities managers are involved in a range of different professional disciplines. It also makes the linkage between the workplace and the impact it can have on its occupants.

The International Facility Management Association (IFMA) defines facility management as:

> *Facility management is a profession that encompasses multiple disciplines to ensure functionality of the built environment by integrating people, place, process and technology.*
>
> IFMA, 2009

This definition, like that of the BIFM, acknowledges the multidiscipline nature of facility management. It makes specific the role of the built environment and even includes technology. However, it lacks the connection to the core business strategy.

The Facility Management Association of Australia (FMA) defines facilities thus:

> *Facilities can be generally defined as buildings, properties and major infrastructure, also referred to within the Facility Management industry as the 'built environment'.*
>
> FMA, 2009

FMA propose that:

> *The primary function of Facility Management (FM) is to manage and maintain the efficient operation of this 'built environment'.*
>
> FMA, 2009

Whilst this definition acknowledges that the built environment needs to be maintained efficiently, it does not link the provision of facility management to the needs of the organization's core business.

All the definitions used for facilities/facility management have their strengths and weaknesses. The essential ingredients of facilities/facility management are:

- it encompasses multidisciplinary activities
- it improves the effectiveness of the core business
- it acknowledges the impact the workplace has on people's productivity

Corporate Real Estate Management (CREM)

Corporate real estate management (CREM) tends to relate to a service provided to an organization whose core business is not real estate (Bon, 1992). The aim of CREM is to contribute to organizational performance by ensuring the real estate portfolio is in alignment with the business strategy. The original ideas associated with CREM tended to relate to the private sector. However, advancement and developments in the public sector mean that CREM is now used in both sectors (Bon, 1992).

A definition of CREM was proposed by Bon:

> *CREM concerns the management of buildings and parcels of land at the disposal of private and public organizations which are not primarily in the real estate business. An organization which occupies space is in the real estate business and needs to manage it properly.*
>
> Bon, 1992, p. 13

The use and ownership of real estate is common to all businesses and yet it has not traditionally been treated in the same way as other company assets. Increasingly companies are now looking at the relationships between real estate and their core business in three key dimensions (Bon, 1994):

- **Financial asset.** This relates to the economic value of the real estate and can be represented as a fixed asset on the company accounts balance sheet. This dimension of assets can be considered to be related to the professional label of asset management.
- **Physical asset.** This relates to the physical provision of land and buildings. The buildings provide a physical envelope in which an organization performs its core business. This can be considered to be related to the professional label of real estate management or property management.
- **Operational asset.** It is the space inside the buildings that has an operational value to an organization. The operational asset relates to the organizational space and service provision. This can be linked to the professional label of facilities management or workplace management.

Whilst the general term that links all these professional services together is corporate real estate management (CREM), some authors, ourselves included, see the increasing emphasis on the three dimensions of asset, and use the term corporate real estate asset management (CREAM).

Skills and Education Requirement

The increasing demands on the CREAM professional mean that to achieve a fully integrated solution a number of professional boundaries have to be collapsed. The

CREAM professional will need to develop business and management skills if they are to effectively support the organization's corporate strategy. In addition it is important that CREAM professionals understand the psychological impacts of the workplace and therefore must also add environmental psychology to their skills development (see Figure 2.1).

Both authors of this book teach CREAM modules, at undergraduate and post-graduate levels, that enable students to acquire this full range of education and skills. The following is a list of expected learning outcomes from our CREAM module:

- read, interpret and understand the language of basic corporate accounts, business plans, and property strategies and extract relevant business and real estate data;
- apply appropriate business and consultancy tools (e.g. SWOT, PEST, Portfolio Matrix Analysis, and Scenario Analysis) to corporate real estate problems;
- build an argument as to how real estate should appear in company accounts and the impact of applying a variety of ownership strategies to corporate accounts and performance;
- extract, construct and critically apply relevant internal performance indicators and benchmarks for business, real estate and mixed indicators (linking business to real estate);

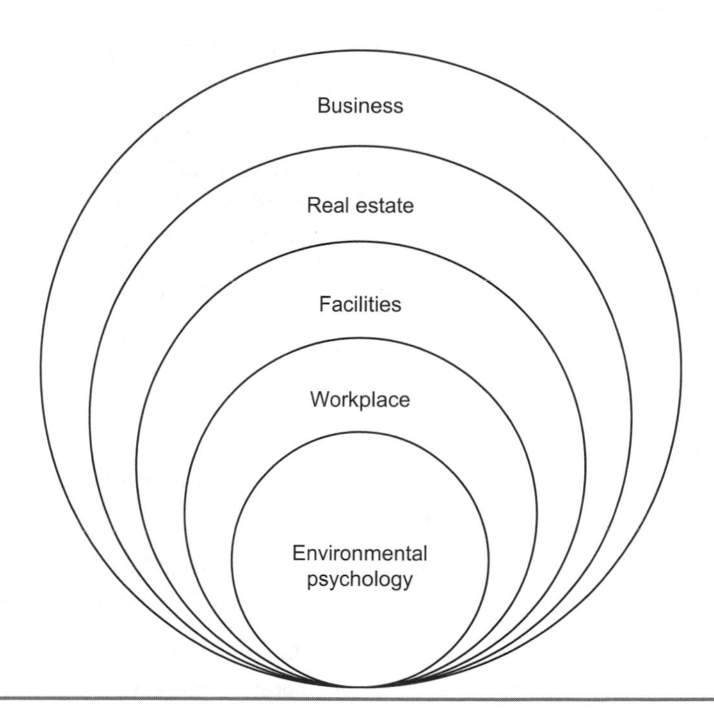

Figure 2.1 Collapsing of professional boundaries

- source, interpret and utilize appropriate external benchmarks to provide externally verified performance analysis;
- identify and be able to critically evaluate the linkages between corporate strategy and real estate strategy and prepare real estate strategies that integrate with and support business objectives;
- identify the key drivers of corporate location decision making;
- demonstrate the capacity to think strategically in relation to complex corporate real estate problems and find integrated solutions;
- apply the research, analysis and tools required to support a successful corporate relocation, including adjacency analysis, space planning, space allocation and design guidance and criteria based building and location assessment;
- utilize financial analysis in corporate real estate problems and solutions;
- appreciate the psychology of space and its role in knowledge management within organizations.

Corporate Strategy

The last section identified the need for the CREAM professional to be able to evaluate, interpret and respond to an organization's corporate strategy. This section will identify the key components of corporate strategy.

Johnson, Scholes & Whittington (2008), in *Exploring Corporate Strategy*, define strategy as follows:

> *Strategy is the direction and scope of an organization over the long term, which achieves advantage in a changing environment through its configuration of resources and competences with the aim of fulfilling stakeholder expectations.*
>
> **Johnson, Scholes & Whittington, 2008, p. 3**

This definition incorporates a number of key concepts relating to strategy:

- the boundaries of the business and where the business is going over a given time period;
- how it intends to use its resources to compete in the marketplace against its competitors;
- constant environmental scanning of the business environment;
- ensuring stakeholder expectations are completely met.

Johnson, Scholes & Whittington (2008) provide the framework for management theory by breaking the management function into three important components:

- **The resources of the organization.** This includes property, which may be a firm's largest asset and one of its highest costs; therefore property needs to be on the strategic agenda.

- **The business environment.** This is changing rapidly and in turn is affecting the way in which property is being used and managed.
- **The direction and scope of an organization.** The nature of an organization, its culture and processes will drive the way in which it uses property. Real estate and facilities must understand the needs of the organization/client and its business strategy to provide a satisfactory real estate/facilities solution.

THE STRATEGIC MANAGEMENT PROCESS

Figure 2.2 illustrates a classic model for the strategic management process (Johnson, Scholes & Whittington, 2008).

The Strategic Position

Understanding the strategic position of an organization involves undertaking a strategic analysis in terms of its external environment, internal resources and competencies, and the expectations and influences of stakeholders.

- **The environment.** Any organization exists in a complex environment that has political, economic, social and technological dimensions. Changes in these dimensions can give rise to opportunities to be exploited or threats to the current nature of an organization. Analytical tools that examine the drivers for change in the environment have been considered in Chapter 1.
- **The resources and competencies of the organization.** It is these components that create an organization's strategic capability. This part of strategic analysis

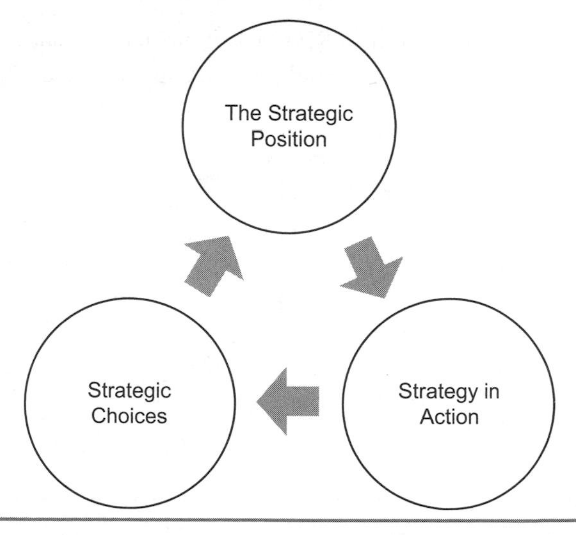

Figure 2.2 The strategic management process. *(Adapted from Johnson, Scholes & Whittington, 2008, with permission)*

examines the internal influences upon the organization: its strengths and weaknesses.

- **Expectations and purpose.** The expectations of different stakeholder groups will also shape the organization and analysis of this should be included in the strategic analysis process.
- **Culture.** Identify the role that culture plays in strategy.

Strategic Choices

Strategic choice involves understanding the underlying principles that guide future strategy and the generation of options for evaluation and selection. The options will be selected upon the basis of organizational mission, culture, market position and resources.

The creation of a CREAM strategy to support the corporate strategy is interdependent on the planning for human resources and information technology (Figure 2.3). If information and communication technology (ICT) is used to enable mobile working then this means that the requirement for organizational space could potentially be reduced.

Achieving this fully integrated approach can bring with it its own challenges.

> *Within most organizational structures practical integration is extremely hard to achieve. These three functions are deemed to require different skills from different people with different backgrounds, all of whom have histories of working in distinctly separate ways.*
>
> Duffy, 2009

The strategic options created by CREAM will include alternative organizational structures of the CREAM department and procurement options of the buildings and

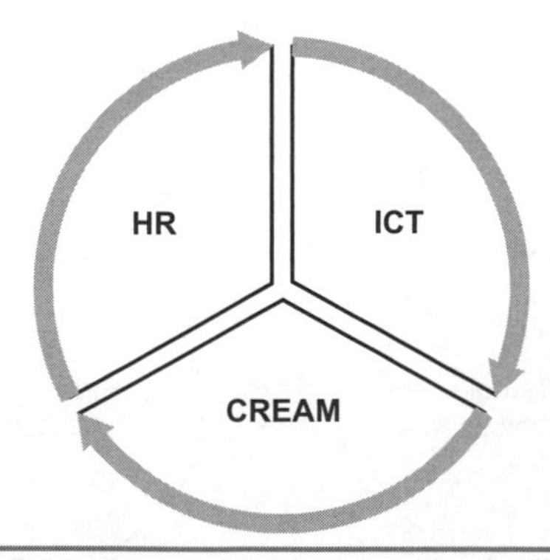

Figure 2.3 An integrated strategic solution

service provision. The procurement options will include issues such as in-house provision versus outsourced provision. The procurement options chosen will have a direct impact on the CREAM organizational structure.

Strategy in Action

The final and perhaps the most difficult part of the process is implementation of the strategy. This may require significant change management programmes, shifts in culture and new working practices, which may be resisted by members of the organization.

The Nokia case study (p. 253) illustrates how an organization implemented and integrated CREAM, human resources and the ICT strategy. ICT was used to enable more flexible mobile working, which ultimately led to a reduction in organizational space requirements. The Nokia case study also illustrates that an essential ingredient to a successful strategic implementation is a well developed change management programme.

Aligning Real Estate Strategy with Corporate Strategy

The corporate real estate an organization uses can have a direct impact on the corporate strategy, and ultimately the organization's performance. Identifying the linkages between corporate strategy and corporate real estate strategy, Roulac (2001) states:

> *A corporate business strategy addresses such critical elements as customers, employees and processors. These elements are profoundly impacted by the environments in which the company does business – the environments in which the enterprise interacts with customers, houses its people and supports its processors. These are elements of corporate property/real estate strategy.*
>
> **Roulac, 2001**

To ensure that these linkages are made, there is a requirement for corporate real estate to be discussed and developed at a strategic level. During the formation of the corporate strategy the corporate real estate manager can provide strategic real estate options that not only support the corporate strategy, but can also act to inform corporate strategy development. The complexities of corporate strategy may require multiple rather than a single real estate strategy (Nourse & Roulac, 1993). Nourse & Roulac (1993) propose eight alternative strategies, which are summarized in Table 2.1.

Nourse & Roulac (1993) identify that irrespective of which real estate strategy an organization adopts, there are a few fundamental considerations that are common to

TABLE 2.1 Alternative Real Estate Strategies

Strategic objective	Real estate strategy
Occupancy cost minimization	This strategy requires that cost effectiveness is maximized
Flexibility	Facilities and workspaces should be provided that can adapt to multiple uses, thereby supporting change in organizational space requirements
Promoting human resources objectives	Acknowledging that the workplace has an impact on employees' satisfaction and productivity
Promote marketing message	The physical buildings and internal working environments can act as a marketing tool for an organization. The workplace could also be used to create brand identity
Promote sales and selling process	Ensure that the locations of the organization's properties are in a location that attracts a high traffic level of customers. The decorations inside the buildings should be attractive to help support and enhance customer sales
Facilitate and control production, operations, service delivery	Ensure facilities are provided in locations that enable linkages to be made with both customers and suppliers
Facilitate managerial process and knowledge work	The physical workspace should be designed to enable and enhance knowledge working
Capture the real estate value creation of business	An example of this strategy would be a major retail outlet acting as an anchor tenant for a shopping centre. Their very presence in the shopping centre can attract footfall and other retail outlets

all strategies. These are considerations that should be addressed during implementation and include (Nourse & Roulac, 1993, pp. 4–5):

- the changing role of real estate in business;
- market conditions concerning relative availability of the quantity, pricing and type of space a company might seek;
- the importance of a special or unique real estate environment to company operations.

Incorporated in the eight alternative real estate strategies are a number of interconnections with other parts of the organization. Clear linkages can be made with marketing and human resource management. It is the interconnected nature of these functions that enables an integrated strategy to be developed.

An integrated corporate real estate strategy can provide a competitive advantage to an organization. Evaluating how organizations can use their corporate real estate to obtain a competitive advantage requires an integrated strategic model (Singer, Bossink & Putte, 2007). The model proposed is literature-based, and combines literature from real estate strategy and competitive strategy.

The corporate real estate strategies adopted include three alternative generic strategies adopted by O'Mara (1999):

- **Incremental strategy.** This is where the real estate strategy evolves on an ad-hoc basis. This ultimately leads to a range of different kinds of properties in the property portfolio.
- **Value-based strategy.** This strategy aims to use the real estate to express the culture and image of the organization. It incorporates the views of the employees and the customers.
- **Standardization strategy.** This strategy concentrates on cost control through standardization.

The competitive advantage strategies used were based on Michael Porter's three generic strategies for sustainable competitive advantage (Porter, 1980):

- **Lowest cost strategy.** This is where the organization aims to achieve the lowest cost product or service to the marketplace.
- **Differentiation strategy.** This strategy is adopted when an organization has something unique in the marketplace. It usually attracts a premium price for a relatively high price.
- **Focus strategy.** This is where an organization targets a distinctive market segment. It is sometimes referred to as targeting a niche market. The focus strategy can be either a low-cost focus strategy or differentiation focus.

Using the generic real estate and competitive advantage strategies, 10 case studies were evaluated to establish the relationship between real estate strategy and competitive strategy (Singer et al., 2007). The results indicate that a standardized real estate strategy supported all of the three generic competitive advantage strategies. The value-based real estate strategy supports a competitive strategy of differentiation and differentiation-focus. However, it does not contribute to a low-cost strategy, or a lowest-cost focus strategy. Clearly the value-based real estate strategy has the potential to emphasize human resources and the marketing aspect of the organization. Finally, the incremental real estate strategy was found not to support any of the three competitive strategies.

In Figure 2.4 we present the classic model for the strategic management process, again with the addition of examples of real estate strategy issues and indicating the chapters in this book where they are discussed in detail.

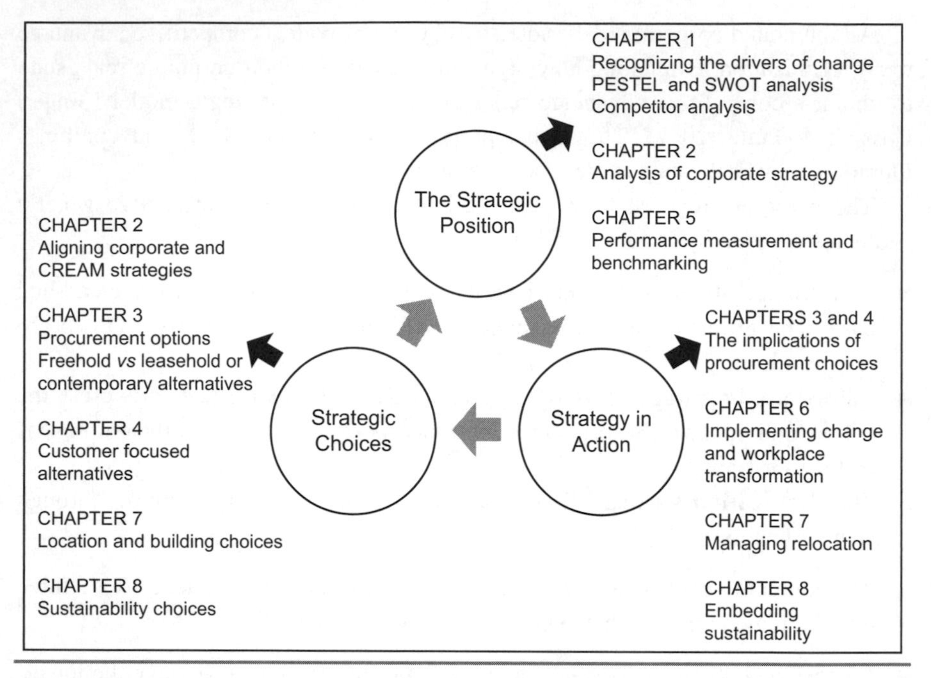

Figure 2.4 The classic strategy model of Johnson, Scholes & Whittington (2008) represented with CREAM attributes

ALIGNING REAL ESTATE STRATEGY WITH ORGANIZATIONAL CULTURE

When considering a real estate strategy for an organization it is important to identify and understand the beliefs and values of the organization. These components are often referred to as the organizational culture. Establishing an understanding of the organizational culture is an integral part of the real estate strategy development.

The Cultural Web

Understanding organizational culture becomes even more important when an organization is undergoing change. However, defining the elements of organizational culture can be elusive. A model that allows a structured evaluation of organizational culture is 'the cultural web' (Johnson, Scholes & Whittington, 2008) (Figure 2.5).

The cultural web enables some of the implicit cultural assumptions to be made explicit. It is only when the organizational cultural assumptions are in the open that a CREAM strategy can be fully developed.

The six elements are:

- **Symbols.** These are the visual cues that are sent out throughout the organization. They can include company logos, the size and quality of someone's office environment, company cars, and the formal or informal dress codes.

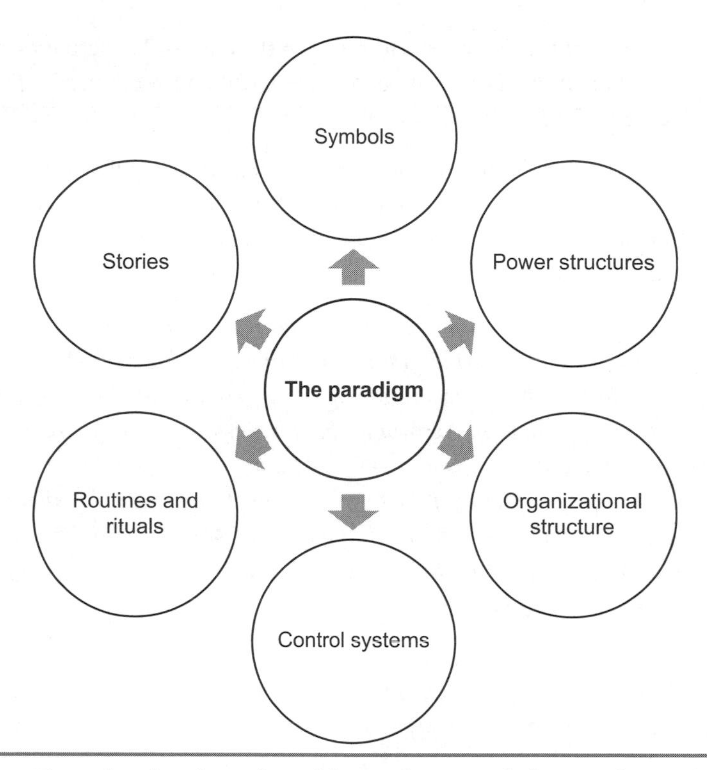

Figure 2.5 The cultural web. *(Adapted from Johnson, Scholes & Whittington, 2008, with permission)*

- **Power structures.** These will be related to the people who make the key strategic decisions of the organization. Determining if these people are located with all other employees or are located as a separate grouping can send clear cultural signals.
- **Organizational structures.** The hierarchy in an organization can send signals as to where you are in the pecking order. A flat hierarchy sends signals that all employees are equal. This then needs to be turned into equitable allocation of space standards.
- **Control systems.** This relates to how the employees are monitored. If the organization adopts a presenteesim culture then this will have a direct impact on the CREAM provision. However, if the organization adopts an output performance measurement system then employees can be more flexible and mobile in their working.
- **Stories.** Stories that are passed from one employee to another can create a context for understanding the organization. A story about a CEO who would rather move a filing cabinet to cover a hole in their office carpet rather than have a new carpet sends clear signals that they are a cost-driven CEO!

- **Rituals and routines.** These tend to relate to the day-to-day activities of people within an organization. They can be very powerful and send clear signals about acceptable behaviour within the organization.

The buildings, space, office layout and furnishings can all act as a physical manifestation of the culture of the organization. When asked how the corporate real estate solution can impact on the organizational culture, Franklin Becker makes the following observations:

> *It conveys through how space is allocated (where are departments located and who gets what space) how individuals and functions are valued. By where people are expected to interact (in an office, in a cafe, in a conference room) and how formal or informal the furniture and its arrangements, cultural values about status and communication.*
>
> *In fact, the entire facility is a form of nonverbal language, silently but often vigorously sending environmental messages about such things as what and who counts and what and who does not; about how formal or informal people are expected to be with each other.*
>
> **Becker, 2008, p. 308**

The CREAM strategy can be used to enhance the current organizational culture, or could be part of a change management programme to change the organizational culture.

Aligning Real Estate Strategy with ICT

A report commissioned by the Royal Institution of Chartered Surveyors (RICS) entitled *Making ICT Work for You*, established the linkages and the potential benefits of information, communication and technology (ICT) to the corporate real estate (CRE) manager (Waller & Thompson, 2009). The report identifies two clear areas where ICT can assist the CRE manager in their role:

- **Managing buildings.** ICT can be used to ensure that the management of space and buildings done by the CRE manager is at an optimum efficiency. A Building Management System (BMS) can integrate a number of building services such as lighting, heating, security, CCTV and alarm systems, access control, energy management and a number of other building services. This fully integrated approach adopts the principles of the 'Intelligent Building' concept. Adopting this approach ensures that optimum control of the building services is achieved, with the benefits of providing energy efficiency gains and a comfortable and productive working environment for its occupants.
- **Managing the interfaces between people and buildings.** This is an area where the CRE manager not only interacts with the ICT manager, but also with the HR

manager. The requirements for knowledge workers to interact and collaborate mean that physical spaces have to be provided which help facilitate these interactions. However, with office workers becoming more mobile, interactions are not always undertaken in the same physical location. The developments in communication technologies and Internet services mean that teams can collaborate virtually. Establishing the balance between the right physical space and virtual space means that the CRE manager and the ICT manager have to work closely together.

The application of ICT enables the CRE manager to monitor and control the resources such as buildings and workspaces at an optimum level. In addition, ICT in the organization allows different types of work process, such as virtual working. The CRE manager has to accommodate changes in working practices, which means a closer working relationship with the ICT and HR management functions.

Adding Value

Private sector organizations tend to be in business to make money. It is their primary aim to generate as much profit as possible so that they can provide the maximum return to their shareholders. Public sector organizations may have a different organizational purpose, but they still have to demonstrate to their stakeholders that their real estate and facilities services provide value for money. Given this context there are increasing demands for non-core activities such as corporate real estate management (CREM) to demonstrate to the organization how it can add value to the organization. However, making the linkages between the corporate strategy and real estate strategy requires an integrated measurement system (Lindholm, Gibler & Kevainen, 2006).

Demonstrating how real estate and FM can add value to an organization requires a shift from the traditional way of thinking. Professionals who work in real estate and FM are used to being asked to cut costs. Organizations tend to see the CREM departments as a cost to the organization (Nourse & Roulac, 1993). Therefore, if a department is defined as a cost centre it is understandable that an organization thinks it can save money by asking that department to reduce costs. The challenge facing the profession is to change the way an organization views the role of real estate and FM. There is a requirement to move from the traditional way of thinking to a new way of thinking. This requires a 'paradigm shift' (Kuhn, 1962). There needs to be a shift from a 'cost reduction paradigm' to a 'value added paradigm' (Haynes, 2007).

If real estate and facilities professionals are to reposition themselves in the eyes of the organization, they need to consider how they can measure their performance in an organizational context. This means that performance measures have to include more than the traditional cost per square metre. The call for new performance measures is supported by Lindholm (2008):

> *Performance measures used in CREM should be identified based on the company's core business goals instead of using traditional accounting measures focusing mainly on cost reductions or capital minimization.*
>
> Lindholm, 2008, p. 344

The traditional measures of CREM tend to be based around measuring efficiency. This approach concentrates on 'doing things right'. However, if measures are to be developed that support the organization then they should be measures relating to effectiveness. This approach leads to 'doing the right thing' for the organization.

The CREAM manager can demonstrate their value-added worth to their organization if they can show that the CREM strategy supports the corporate strategy. A strategic framework that illustrates the linkages between corporate real estate management and the business strategy is illustrated in Figure 2.6 (Lindholm et al., 2006).

The strategic framework in Figure 2.6 starts with the premise that an organization's primary aim is to maximize wealth of their shareholders. The generation of wealth of the shareholders is achieved by the development of the business strategy. The CREM role is to support the business strategy and add value back to the

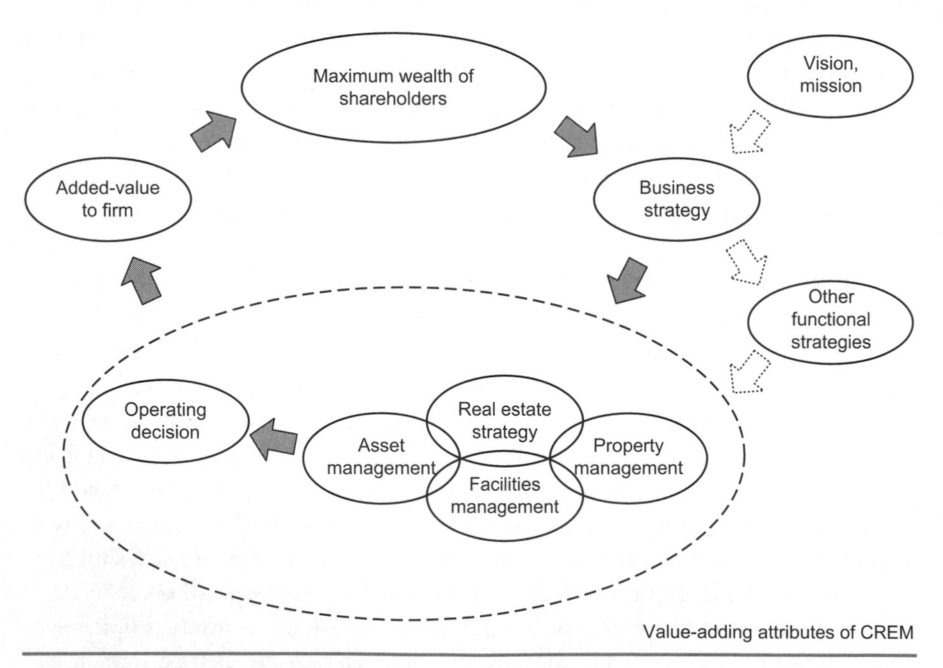

Figure 2.6 CREM as a part of the firm's strategic framework. *(From Lindholm et al., 2006; reproduced with permission)*

organization therefore ensuring that the shareholders receive maximum return on their investments.

Understanding the organization's business strategy is the start point for the creation of a supporting real estate strategy. Business strategies can vary depending on such things as the competition in the marketplace, the state of the business environment, competitors and the expectations of customers and clients. Consequently there will be a range of different real estate strategies to support the range of possible business strategies.

Real estate strategies that could be adopted for the business strategies of revenue growth or profitability growth are presented by Lindholm et al. (2006) and are set out below.

- Real estate strategies that support the core business strategy of revenue growth could include:
  - Increase the value of assets. This would include issues relating to the lease or purchase of the organization's properties.
  - Promote marketing and sales. Selecting the appropriate location of property to attract customers.
  - Increase innovations. The creation of usable and innovative workplaces.
  - Increase employee satisfaction. Selecting the appropriate location of property to attract employees.
- Real estate strategies that support the core business strategy of profitability growth could include:
  - Increase employee satisfaction. Provide environments that support employees with the desired amenities.
  - Increase productivity. This can be achieved by creating a new working environment that enhances productivity.
  - Increased flexibility. Create flexible workplaces and ensure the balance between owning and leasing is achieved.
  - Reduce costs. Ensure efficient cost management systems are in place.

Once the real estate strategy that most aligns to the business strategy is decided upon, consideration can be given to the implementation and measurement of the strategy. This will require a range of performance metrics that are uniquely connected to the choice of real estate strategy.

> *Establishing tools and indicators dedicated to measuring the added value of real estate to overall strategic corporate objectives is therefore eminent for corporate real estate departments in order to be considered a fully fledged participant in the strategic decision-making process at the top of management.*
>
> **Scheffer, Singer & Meerwijk, 2006, p. 197**

In an attempt to raise the strategic profile of real estate, Scheffer et al. (2006) propose a measurement tool that enables corporate real estate executives to specifically identify which elements of corporate real estate enhance corporate strategy. The measurement tool proposed integrates the frameworks of Nourse & Roulac (1993) and De Jonge (1996) to provide an integrated model, as can be seen in Figure 2.7.

The driving forces help define direction of the business and relate to the guiding forces that inform the organization's corporate strategy. Nourse & Roulac (1993) identify nine potential driving forces.

De Jonge (1996) identifies seven potential elements of added value. The measurement tool proposed by Scheffer et al. (2006), as can be seen in Figure 2.7, establishes for each driving force the relevant added values components that should be part of the real estate strategy.

If an organization's main driving force is its products then the corporate strategy would define the business by the products it produces. Interpreting the results in Figure 2.7, it can be seen that adopting this driving force the real estate added values elements would be:

- **Increasing productivity.** This could be created by innovations in the workplace and the use of the working environment as a way of attracting and retaining the organization's human capital.

		Added values						
		Productivity	Costs	Risk	Value	Flexibility	Culture	PR and marketing
Driving forces	Products	✓	✓	✓				
	Market			✓		✓		✓
	Technology	✓		✓		✓	✓	
	Production	✓		✓				
	Resources	✓		✓	✓			
	Distribution			✓				✓
	Sales		✓	✓				✓
	Growth			✓				✓
	Profit		✓	✓	✓			

Figure 2.7 Relationship between driving forces and added value. *(From Scheffer et al., 2006; reproduced by permission of Emerald)*

- **Cost reduction.** A close monitoring and control of workplace and accommodation costs.
- **Risk control.** Ensure that the real estate portfolio is as flexible as possible so that it can adapt and respond to the changing business requirements.

It should be acknowledged that risk control is a common theme throughout all the proposed driving forces. This indicates that the added value element of risk control needs to be embedded in all real estate strategy proposals (Scheffer et al., 2006).

Adding Value through the Human Asset

Whilst, in most organizations, that the two most important assets are its people and its real estate is well understood, the linkage between the two is not (Becker, 1990). With staff cost in the region of 70–80% and real estate costs approximately 20%, a relatively small increase in the productivity of the employees is greatly more beneficial than a small reduction in real estate costs (Weatherhead, 1997).

The Commission for Architecture and the Built Environment (CABE) established from their own research similar costing ratios.

> *Looking at the discounted present value of developing, owning and operating a typical office building over 25 years on a traditional occupational lease, this shows that, excluding land, 6.5% of the total goes on construction costs; 8.5% goes on furnishings, maintaining and operating the facility; and, dramatically, the balance of 85% goes on the salary costs of the occupants.*
>
> CABE, 2005, p. 9

Looking at the life cycle costing of a typical office building, CABE established that 85% of total life cycle costs were related to human resources costs. In comparison, only 8.5% of the other costs could be attributed to real estate and facilities costs.

This leveraging argument highlights that whilst real estate and facilities managers may be constantly aiming to reduce cost, the greatest gains are to be made if consideration is shifted to improving organizational productivity (Haynes, 2008). This leveraging approach can be argued the opposite way; mismatching people with their work environments could have a significant impact on overall organization performance (Mawson, 2002; Haynes, 2008).

The same 'leveraging' argument is adopted by Becker and Pearce (2003), who propose an integrated cost model. The model consists of both corporate real estate and human resource factors. They call their model the Cornell Balanced Real Estate Assessment model, COBRA, which includes the three main variables: measures of productivity, human resources costs and real estate costs. Together the three variables in the model enable organizations to make strategic real estate decisions.

> *HR impacts can be highly significant, and if incorporated into a single model might lead to recommendations very different from those based only on the direct real estate costs.*
>
> Becker & Pearce, 2003, p. 233

A typical example would be the evaluation of a new capital build. If the choice was between a basic development or one of a higher standard, and subsequent cost, the costing model would predict the appropriate rise in employee productivity required to pay for the more expensive option.

The benefits of linking the corporate real estate strategy with the human resource strategy can be summarized as:

> *For organizations that plan, design, and manage their physical facilities well, there are many strong connections to HR and OD (Organizational Development) issues and goals: a great facility can help attract and retain staff; can motivate them and increase job satisfaction; can help build teamwork and collaboration; can signal changes in culture and key corporate values.*
>
> Becker, 2008, p. 308

The workspace provided for employees to work, interact, collaborate and rest can be the linking component between real estate and human resources. The type and quality of space can impact on many aspects of employees' working life. Identifying and creating spaces, which are particularly important to 'high value' employees, means that an organization can attract and retain the right kind of people (Martin & Black, 2006).

The space created may also be used to enhance the health and well-being of its occupants. When considering a new workplace design there are a number of legislative requirements that need to be considered relating to health and safety. Whilst safety at work has long been a central issue, there is also a need to consider the potential impact the working environment can have on the health of its occupants. Placing plants in the workplace can potentially remove indoor air pollutants and consequently improve the air quality; however the psychological benefit of plants may also have a role to play (Smith & Pitt, 2009).

Adding Value through Brand Identity

The brand of an organization is more than the marketing materials and the company logo; it has to reflect the whole ethos of the organization. It is this ethos, or brand identity, that helps an organization differentiate itself in the market place.

An organization can use its real estate and facilities to communicate clearly to employees and customers its brand identity. The location of an organization's buildings, their external and internal design and appearance all send signals about the company culture and brand.

A successful brand will be fully integrated into the business: it will reflect the culture of the workplace, the values of the organization, the perceptions of customers and clients, even the nature of the business's assets.

McNestrie, 2009

Branding is strongly associated with the consumer and retail sector. However, it can perform just as important a role in office working environments. The branding of office environments can clearly reflect the culture of an organization, or corporate DNA. If the real estate and facilities professionals are not consciously managing the brand of a company, then by default they may be sending negative signals to their customers and employees.

Branding can add value to an organization during times of mergers and acquisitions. When two organizations merge, or an organization is taken over by another, distinctly different brands need to be brought together as one. An integral part of the new corporate strategy would be the rebranding of the newly created organization. The real estate and facilities departments can be an integral part in this rebranding process.

When Thomson, a North American business that provided information to professionals, acquired Reuters, a business-to-business media company, there was a clear need to re-brand the newly formed organization. The new company was called Thomson Reuters and had a new business aim of becoming 'a business Google for the professional industry' (McNestrie, 2009). The new business had to be seen as more than two businesses bolted together. However, both Thomson and Reuters had clear brand identities and whilst the value of these brand assets had to be maintained, there was an opportunity to demonstrate the enhanced services the new organization could offer (McNestrie, 2009).

The rebranding of Thomson Reuters had to be implemented in an organized and coordinated way as it involved 260 locations in 56 countries (McNestrie, 2009). The strategy adopted took a 'follow the sun' approach, which started in Beijing and moved east during the day. The rebranding reached Canary Wharf at 4 am, where new signage was erected.

From when people entered the office, they felt they were working for a new organization.

McNestrie, 2009, p. 49

Establishing a new Thomson Reuters brand had clear business objectives, however it was also important to send clear signals, to the employees of the organization, about the new company ethos. Acquisitions and mergers can be an unsettling time for employees, but the rebranding process can make them feel part of something new.

CREAM Strategic Alignment Model

The alignment of the CREAM strategy to the corporate strategy with the aim of adding value and enhancing organizational performance is the focus of much research. Making the connections between workplace design and occupier productivity is a complex research area that requires innovative methodological research design (Haynes, 2009). This 'holy grail' is yet to be fully researched and validated but on-going work by the authors is seeking to quantify the true impact of real estate to business by examining the benefits of optimal alignment between, position, purpose, place, paradigm, processes and people to produce performance and productivity (Haynes, 2008).

This research has attempted to evaluate how well the office environment supports the office occupier in their work processes. Underpinning this research is the proposition that office occupiers have *'connectivity'* with their office environment. This connectivity is both physical and behavioural and collectively can be termed *'workplace connectivity'*. It is proposed that the alignment of the office environment (place) with the work processes provides increased workplace connectivity and productivity.

To develop this area of research further, it is suggested that there is a need for additional components to be included into a theoretical framework. A proposed theoretical framework can be seen in Figure 2.8 (Haynes, 2008).

It is proposed that a high performance and productive workplace can be created when position, purpose, place, paradigm, process and people are all in alignment. The 8 Ps are as follows:

- **Position.** This component relates to the position of the organization in the business environment relative to its competitors. In addition this component can also relate to the position of the organization in the business cycle. An expanding organization has different requirements to a restructuring or contracting organization.
- **Purpose.** This relates to the business aims and objectives. This includes the company mission or vision statement and aims to establish the future direction of the business.
- **Place.** This has external and internal components. The external component relates to where the office building is physically located. The internal component relates to the building and office environment layout created.

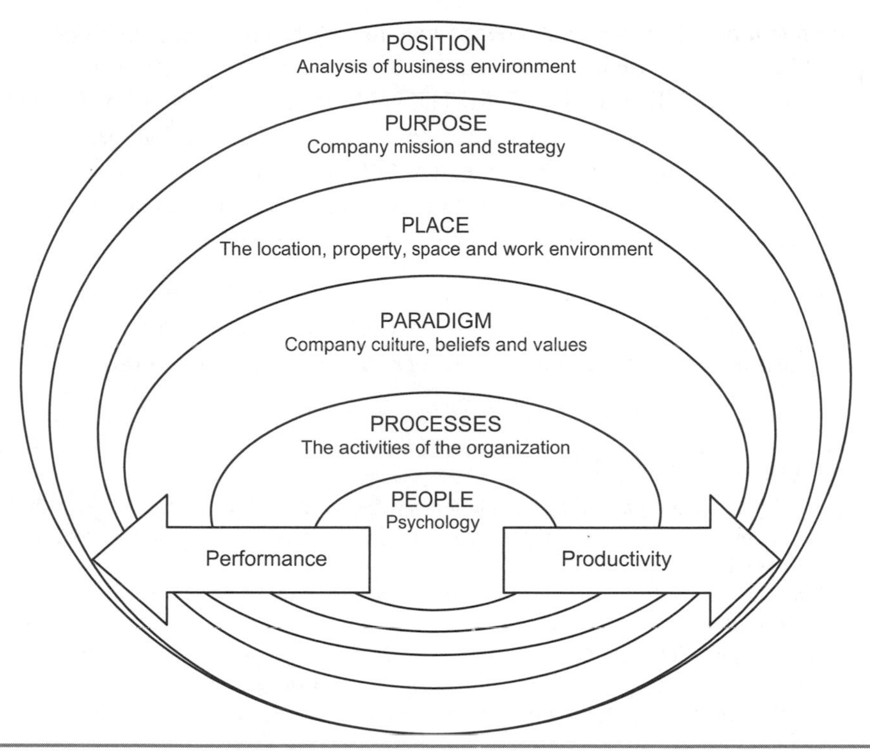

Figure 2.8 CREAM strategic alignment model. *(From Haynes, 2008; reproduced with permission)*

- **Paradigm.** This relates to the organizational culture or *'organizational DNA'*. It is important to establish how an organization actually works and understand its beliefs and values. It is also important to establish micro cultures, the way that people actually work in the office environment; this could be considered to be the *'workplace DNA'*.
- **Processes.** The activities of an organization have a number of different levels. At one level this includes the interrelations between the different departments in an organization, and at another level this relates to the specific work processes undertaken by the individuals in an office environment.
- **People.** Understanding the office environment from the occupier perspective is a central principle to the alignment model. This could be developed to include personality types and team role types.
- **Productivity.** This could include individual and team-based productivity measures. Productivity measures maybe directly linked to business performance, self-assessed measures or a combination of the two.

- **Performance.** This could include the traditional efficiency measures of property and facilities performance, but would also include hybrid measures of business and property/facilities performance. This could include revenue/m^2 or profit/m^2. In addition to evaluating *'effective space'* it is suggested that performance measures could be developed to measure *'affective space'*. Since we experience office environments with all our senses, it seems appropriate to develop means of assessing how the office environment affects those senses.

Since human resources and real estate are usually the two main assets of an organization, the proposed theoretical framework attempts to capture the need to achieve *'asset alignment'* leading to *'asset productivity'* (Haynes, 2008).

Summary Checklist

1. Identify for your organization the relationships between asset management, real estate management, property management, facilities management and workplace management.
2. Establish how CREAM integrates into the corporate strategic development of the organization.
3. Identify the beliefs and values of the organizations using cultural web analysis.
4. Ensure that ICT is fully utilized with regards to the management of CREAM and monitor for any potential implications that may change the future delivery of the business.
5. Establish a strategic relationship with the human resources department so that the people dimension can be integrated into future strategic proposals.
6. Identify the organization's branding strategy. Then integrate into the CREAM strategy thereby ensuring that workplace design reflects the brand identify of the organization.
7. Finally, check that all components of the CREAM strategic alignment model are in alignment, thereby ensuring that the organization receives a strategic solution from CREAM.

References

Becker, F. (2008). Talking Heads. *Journal of Corporate Real Estate, 10*(4), 308–310.

Becker, F., & Pearce, A. (2003). Considering corporate real estate and human resource factors in an integrated cost model. *Journal of Corporate Real Estate, 5*(3), 221–241.

BIFM. (2009). *Facilities management introduction.* Retrieved from British Institute of Facilities Management. http://www.bifm.org.uk/bifm/about/facilities. Accessed 30.07.09.

Bon, R. (1992). Corporate Real Estate Management. *Facilities, 10*(12), 13–17.

Bon, R. (1994). Ten Principles of Corporate Real Estate. *Facilities, 12*(5), 9–10.

Duffy, F. (2009). Talking Heads. *Journal of Corporate Real Estate, 11*(1).

FMA. (2009). *What is Facility Management?* Retrieved from Facility Management Association of Australia (FMA). http://www.fma.com.au/cms/index.php?option=content&task=view&id=45. Accessed 30.07.09.

Haynes, B. P. (2007). Office productivity: a shift from cost reduction to human contribution. *Facilities, 25* (11/12), 452–462.

Haynes, B. P. (2008). Impact of workplace connectivity on office productivity. *Journal of Corporate Real Estate, 10*(4), 286–302.

Haynes, B. P. (2009). Research design or the measurement of perceived office productivity. *Intelligent Buildings International, 1*(3), 169–183.

IFMA. (2009). *FM Definitions.* Retrieved from International Facility Management Association (IFMA). http://www.ifma.org/what_is_fm/fm_definitions.cfm. Accessed 30.07.09.

Johnson, G., Scholes, K., & Whittington, R. (2008). *Exploring corporate strategy text and cases* (Eighth Edition). Harlow: Pearson Education Limited.

Jones, K., & White, A. D. (2008). *RICS Public Sector Asset Management Guidelines: A guide to best practice.* London: RICS.

De Jonge, H. (1996). Toegevoegde Waarde Van Concernhuisvesting. *Paper presented at the NSC-Conference on 15th October, 1996.*

Kuhn, T. (1962). *The structure of scientific revolutions.* Chicago, IL: The University of Chicago Press.

Lindholm, A.-L. (2008). A constructive study on creating core business relevant CREM strategy and performance measures. *Facilities, 26*(7/8), 343–358.

Lindholm, A.-L., Gibler, K., & Kevainen, K. I. (2006). Modeling the Value-Adding Attributes of Real Estate to the Wealth Maximization of the Firm. *Journal of Real Estate Research, 28*(4), 445–474.

Martin, P., & Black, R. (2006). Corporate real estate as a human resource management tool. *Journal of Corporate Real Estate, 8*(2), 52–61.

Mawson, A. (2002). *The workplace and its impact on productivity.* London: Advanced Workplace Associates.

McNestrie, A. (2009). Strike up the brand. *FM World.* (47–49) (February).

Nourse, H., & Roulac, S. (1993). Linking real estate decision to corporate stategy. *Journal of Real Estate Research, 8*(4), 475–494.

O'Mara, M. (1999). *Strategy and place: Managing corporate real esatate and facilites for competitive advantage.* New York, NY: The Free Press.

Porter, M. (1980). *Competitive strategy.* New York, NY: The Free Press.

Roulac, S. (2001). Corporate Property Strategy is Integral to Corporate Business Strategy. *Journal of Real Estate Research, 22*(1/2), 129–151.

Scheffer, J., Singer, B., & Meerwijk, M. V. (2006). Enhancing the contribution of corporate real estate to corporate strategy. *Journal of Corporate Real Estate, 8*(4), 188–197.

Singer, B. P., Bossink, B. A., & Putte, H. J. (2007). Corporate real estate and competitive strategy. *Journal of Corporate Real Estate, 9*(1), 25–38.

Smith, A., & Pitt, M. (2009). Sustainable workplaces: improving staff health and well-being using plants. *Journal of Corporate Real Estate, 11*(1), 52–63.

Waller, A., & Thompson, B. (2009). *Property in the economy: Making ICT work for you.* London: RICS.

Weatherhead, M. (1997). *Real estate in corporate strategy.* New York, NY: Palgrave.

Chapter 3

Strategic Real Estate Procurement Options

Introduction

One of the most important strategic options for an organization is the selection of appropriate procurement options for the organization's real estate portfolio. However, the drivers of change identified in Chapter 1 are making it more difficult to co-ordinate real estate ownership options with the volatility of demand and utilization created by contemporary management practices and economic uncertainty. In this chapter we examine both traditional procurement options, including freehold and leasehold property, and more contemporary options, including flexible leasing products from Landflex and virtual solutions from MWB and 'Regus'. We also consider the implications of the procurement decision, including a brief introduction to the proposed significant changes to the financial reporting of leases by the International Accounting Standards Board.

The Freehold *vs.* Leasehold Debate

We start with a discussion between the classic choices of freehold and leasehold property. Owning freehold real estate has been the historic choice for many organizations. The reasons behind this may have been:

- **cultural** – driven by the desire to have absolute control of their property assets;
- **financial** – driven by the desire to accumulate capital growth;
- **business** – specialist businesses may require bespoke premises that would be difficult to procure under a standard lease.

Ownership of real estate, especially in the UK, confers stability and robustness and it is sometimes difficult to break the *'corporate DNA'* of an organization owing to its culture and management traditions. In addition, through mergers and

Corporate Real Estate Asset Management. ISBN: 978-0-7282-0573-4

acquisitions companies may inherit property that may not fit with an organization's strategy. Constant review of the assets, the distribution of procurement types and the fit of the real estate to the needs of the business is therefore essential.

In addition, an organization has to make some fundamental choices:

- does it want to use property as an asset, with potential for capital growth, and as collateral for future borrowing, the problem being that the capital tied up in the property assets may not be working as hard as capital that could be used in the development of the business; or
- does it need to be flexible and agile, keep capital off the balance sheet and embedded in the core business?

The main advantages and disadvantages of freehold and leasehold occupation are summarized in Table 3.1:

For modern businesses there may be compelling reasons to procure corporate real estate by a mixture of both freehold and leasehold property, within an occupational portfolio. The diversity that this offers has been the subject of previous discussion, including Weatherhead (1997), and we represent and update this in Table 3.2, in terms of strategic and financial issues.

Marks & Spencer (M&S) provides a good example of a company exposed to the advantages and disadvantages of freehold ownership. Writing in the *Harvard Business Review* (July/August 1995), Collins & Montgomery point to M&S as having strategic advantage through their high proportion of freehold property. They indicate that their occupancy costs were 1% compared to a UK industry average of 3–9%. However, in 1999 this exposure to freehold ownership made them vulnerable to a hostile takeover with investors looking to the real estate assets rather than the value of the, then struggling, brand. Subsequently, M&S undertook significant restructuring involving sale-and-leasebacks and a £331 million bond securitized on the rental income from part of their portfolio.

Reasons for Companies Leasing Their Property

The most significant empirical research carried out into the decision-making between freehold and leasehold procurement options in the UK was undertaken by Lasfer, sponsored by Donaldsons and published in 2007. The research examined all quoted companies in the UK between 1989 and 2002.

The research set out to test the main arguments for leasing, identified as:

- to finance growth opportunities;
- have a lower level of debt when leasing is reported as off balance sheet financing;
- obtain lower rental costs where capital allowances are passed on by landlords; and
- increase their efficiency as they treat property as a cost asset and manage it more effectively.

TABLE 3.1 Advantages and Disadvantages of Freehold and Leasehold Procurement

Freehold	Leasehold
Main Advantages	
Provides high degrees of control and security, ability to brand, manage and adapt, especially compared to a multi let building	Provides flexibility of occupation to adapt to changing business cycles and fortunes, especially through short leases and break clauses
Provides a significant fixed asset for future funding/collateral opportunity, especially if not subject to any debt	Reduces fixed assets on the balance sheet which may have positive impacts upon accounting and financial reporting
May provide investment returns in terms of capital growth and in periods of inflation a hedging opportunity	Provides opportunities, especially in a pro-tenant market, for significant rent-free periods and fitting out allowances, especially to quality tenants
Provides security, continuity and stability both physically and psychologically	Provides opportunities for significant additional customer services which may be supplied cheaper than an organization providing them in-house – e.g. meeting rooms, reception etc. Landlords may be able to procure energy and other services across their whole portfolio cheaper than an individual owner and pass on these savings to the lessee
Costs are predictable and for example, if using a long term fixed mortgage, stable and known throughout the ownership period.	Provides opportunities for partnering, clustering and networking with other businesses
Main Disadvantages	
Returns may not exceed those of capital invested in the business, thereby creating an adverse opportunity cost	Rental is a significant operating cost which is unpredictable and may spiral in certain market conditions and be difficult to forecast over an extended period
Capital gains may be eroded by property obsolescence, adverse change in a locations status or other real estate risks	Lack of control and in some cases onerous leases which require consent for every minor alteration to the building
May depress key financial ratios when compared to similar business with less fixed assets on their balance sheets	Inability to tailor a building to the personal requirements of an organization for example through external signage or branding
Significant management costs may be incurred in providing facilities and building management services to keep the building operational. If these are done in-house they may be a distraction from the core business and a significant management burden	Sharing floors or common areas may impose unacceptable security or other risks

TABLE 3.2 Strategic and Financial Aspects of Real Estate Procurement

Freehold	Leasehold
Strategic Issues	
If property has additional land, opportunity for future expansion	Less risk of being locked into an unsuitable or obsolete building
Self-contained, controlled and managed to an organization's own specifications and expectations	Opportunity to test a locality, landlord or building type without a major commitment
Longevity and stability sends a strong cultural message of permanence	Allows project and short-term team working and operations
May be appropriate if investment in plant and machinery has a long life cycle and/or is difficult to relocate	Ability to purchase additional services especially in a serviced office (e.g. Regus or MWB) or customer focused landlord (e.g. Bruntwood)
Flexibility to dispose of, but only in positive market conditions	May allow occupation within a locality or building occupied by complimentary tenants or users leading to economies of scale or partnering
Could allow for inclusion of specific facilities to promote staff or customer welfare which may not be permissible in a lease situation	
May allow an organization to locate where no suitable leasehold property exists	
Financial Issues	
Isolates an organization from cash flow shocks at rent review or through service charges	Demands less capital which can be used in the business
Avoids onerous lease conditions and long term commitments	Capital may be better utilized, giving a higher return within the business
Greater control over management costs and maintenance	Does not place a large amount of fixed asset in the balance sheet which may depress key financial ratios when compared to competitors
Allows greater protection of expensive investment in plant, machinery and facilities	Limits the proportion of non-liquid capital assets
May attract grants or other incentives	May give more freedom of choice, in terms of location, landlord, building design and configuration
Potential for expansion or alternative use development in the future which may create added value	

TABLE 3.2 Strategic and Financial Aspects of Real Estate Procurement – cont'd

Freehold	Leasehold
In times of low interest rates and high rentals may be attractive alternative to leasing	
May provide opportunities for inclusion in an organization's or director's pension fund	
May attract capital allowances and other tax incentives	

The core findings of the research were:

- the decision to lease property is driven by strategic business considerations, in particular to reduce debt, to finance growth and through the passing on of tax allowances to conserve cash;
- leasing does promote more efficient use of real estate assets;
- companies that report 100% leasing were found to generate higher total returns to their shareholders than those who reported only freehold real estate; and
- a suggested optimal level of leasing of 65% leasing to 35% ownership maximizes the returns to shareholders by balancing the above advantages of leasing with the loss of collateral and increased bankruptcy risk associated with too much real estate ownership.

The research also reported other interesting findings including:

- a significant increase in companies leasing real estate in the UK;
- the propensity to lease varies by industry, the highest being in information technology companies and the lowest in automobile, household goods and textiles;
- the propensity to lease varies by the scale of the organization, with larger companies leasing more real estate;
- liquidity ratios are higher in companies that lease property, as leasing appears to promote better cash conversion and lower inventory;
- the corporation tax liability of companies that lease is significantly lower suggesting that companies lease property because they cannot claim tax allowances.

In summary, the basic freehold *vs.* leasehold decision has a number of strategic, financial and operational factors which will vary from company to company. It will be influenced by the property market cycle, tax provisions and accounting practice. Property incurs significant transaction costs and therefore the decision process is subject to long lags between say a decision to buy and completion of the transaction, which makes it imperfect. Weatherhead (1997) sets out a very simple decision-making table, which we have updated to reflect further research and contemporary practice and is set out in Table 3.3.

TABLE 3.3 Freehold *vs.* Leasehold Decision-making Matrix

OWN *if*	Business factor	RENT *if*
Stable	Operation	Changing
Return on real estate > than return on business	Financial management	Return on real estate < than return on business
Unique	Accommodation needs	Ordinary
Heavy	Branding	Light
Specialized and expensive	Plant and machinery	Non-specialized
Need for control and security	Management	Flexibility required and is team- or project-based
High	Rental inflation	Low
High	General inflation	Low
Available	Availability of capital	Restricted
Low	Cost of capital	High
Under 35% freehold	Current portfolio	Over 35% freehold
ABLE to claim capital allowances against Corporation Tax	Tax allowances	NOT ABLE to claim capital allowances against Corporation Tax

The Financial Argument

We have indicated above that there are a series of corporate real estate issues around the buy/lease decision. If, however, the decision-making analysis points to equal positives and negatives then a cash flow analysis may be useful to examine the net present values of the available procurement options.

In simple terms a cash flow can be constructed over, say, a 10-year period which examines the inflows and outflows associated with a leasehold and freehold decision. In the freehold scenario this will assume the payment of a deposit with a fixed rate mortgage paid over 10 years and potentially an assumption of capital growth creating equity at the end of the holding period. For the leasehold a series of outflows representing rent, service charges and taxes will be entered into the cash flow. Whilst in most cases mortgage costs will be higher than those for leasing, the increase in capital value may potentially offset the extra annual costs in the 10-year cash flow. Of course, the value will be determined by market conditions at the time and it may be difficult to predict this component of the analysis. These assumptions are set out in Table 3.4.

We have kept the example in Table 3.4 as simple as possible to illustrate the technique; of course in practice greater consideration of the relative costs of service

TABLE 3.4 Assumptions for the Comparison of a Freehold or Leasehold Procurement Decision in a Simple Buy/Lease Model

Building A: Leasing option	Building B: Purchasing option
2500 sq ft lettable area	2500 sq ft lettable area
Initial rental: £100,000 p.a. Assumption that rent will increase to £130,000 at the start of year 6	Purchase price: £2,000,000 Financing available on an interest only basis at 7% p.a. with a loan to value ratio of 70%. Mortgage costs = 7% of £1,400,000 p.a. = £98,000 p.a. Opportunity cost of the equity contribution is equal to the return on the capital invested in the business of 10% (10% of 600,000 = 60,000 p.a.) Assumption that the capital value will rise to £2,750,000 in Year 10
Acquisition costs: £2,000	Acquisition costs: £10,000 Disposal costs: £10,000
Service charge assumed to equal planned maintenance and sinking fund expended by owner – therefore ignored for simplicity.	Service charge assumed to equal planned maintenance and sinking fund expended by owner – therefore ignored for simplicity.
Dilapidation costs assumed to be equal to preparation for sale costs	Dilapidation costs assumed to be equal to preparation for sale costs

charge, maintenance, dilapidations, management and other costs associated with the procurement would require evaluation.

The leasehold analysis calculates the discounted cash flow for the outgoings of rent plus the initial acquisition costs and is set out in Figure 3.1.

The freehold analysis calculates the mortgage costs and opportunity costs of capital and reflects the projected capital value increase when the property is sold at the end of the 10-year holding period. The model assumes that when the property is sold for £2,750,000 the capital (both loan and equity) is repaid from the proceeds, as is set out in Figure 3.2.

It can be seen in this example that the freehold option is marginally the better financial option. However, this analysis is highly sensitive to changes in the interest rates, projected capital growth, cost of equity (opportunity costs) and rental inflation. We would advocate sensitivity testing of these variables to analyse a series of potential outcomes determined by different scenarios of the future.

There are a number of companies offering software solutions to this decision-making problem which provide detailed scenario planning and the ability to calculate break-even rents and capital values to show which decision is preferable.

Leasehold option		*Costs*			
Year	Acquisition/Disposal	Rent	Total costs	Present value @ 8%	Discounted outgoings
1	−2,000	−100,000	−102,000	1.0000	−102,000
2		−100,000	−100,000	0.9259	−92,590
3		−100,000	−100,000	0.8573	−85,730
4		−100,000	−100,000	0.7938	−79,380
5		−100,000	−100,000	0.7350	−73,500
6		−130,000	−130,000	0.6806	−88,478
7		−130,000	−130,000	0.6302	−81,926
8		−130,000	−130,000	0.5835	−75,855
9		−130,000	−130,000	0.5406	−70,278
10		−130,000	−130,000	0.5002	−65,026
			TOTAL COSTS (−NPV) =		**−814,763**

Figure 3.1 Discounted cash flow analysis of the leasehold option

Getting the Right Balance of Procurement

Lease Length

For an occupier lease length is a further strategic consideration once a decision to lease has been made. As discussed above, the availability of lease length will be determined by market conditions and the relative negotiating power of landlords and

Freehold option	*Costs and capital receipts*						
Year	Acquisition/ Disposal	Mortgage interest	Opportunity cost of equity	Capital appreciation	Total costs	Present value @ 8%	Discounted outgoings
1	−10,000	−98,000	−60,000		−168,000	1.0000	−168,000
2		−98,000	−60,000		−158,000	0.9259	−146,292
3		−98,000	−60,000		−158,000	0.8573	−135,453
4		−98,000	−60,000		−158,000	0.7938	−125,420
5		−98,000	−60,000		−158,000	0.7350	−116,130
6		−98,000	−60,000		−158,000	0.6806	−107,535
7		−98,000	−60,000		−158,000	0.6302	−99,572
8		−98,000	−60,000		−158,000	0.5835	−92,193
9		−98,000	−60,000		−158,000	0.5406	−85,415
10	−10,000	−98,000	−60,000	750,000	582,000	0.5002	291,116
					TOTAL COSTS (−NPV) =		**−784,894**

Figure 3.2 Discounted cash flow analysis of the freehold option

tenants, funding arrangements and leverage, especially on new buildings and the customer focus perspective of the landlord.

In simple terms the shorter the lease the greater the business flexibility. If a business has a very short life cycle or is subject to extreme volatility it should consider a short lease or even a 'pay as you go' style operation, which we discuss later in the chapter. However, many businesses invest considerable amounts of money in fitting out, furnishing and branding their office or retail space and it may not be possible to recover those costs over the length of a very short lease.

An agreed lease length will therefore be determined by both supply and availability and by the particular demand characteristics of an individual occupier.

A very interesting and provocative viewpoint on lease length has recently been revisited by Dr Peter Linneman, writing in *The Leader*, the Journal of CoreNet Global, although his arguments first appeared in a paper in 1998. Linneman believes that the real cost of going long, either by ownership or lease in real estate, is that you are likely to be 'stuck' with something you don't want and that cost is very hard to calculate. His argument is that if you try to calculate with precision the cost of going for a long lease it would be very difficult because there are so many different scenarios where the space may no longer be required.

However, the costs of going short, taking a short-term lease and finding out you were wrong, may be far less. He suggests that having a 5-year lease and then needing it for 10 may result in a few extra dollars a square foot, which is both tangible and manageable. He suggests this is a much more comfortable cost than committing to say a 15-year lease and not needing it after 5 years. In a difficult market, with low inflation, sub-letting may be difficult and/or economically unrewarding.

Linneman (2007) therefore advocates short leases with the relatively predictable cost of a mistake and this resonates with the increasing flexibility and volatility of the business environment which we explored in Chapter 1.

The Service Flexibility Continuum

There is a continuing movement to more flexible and more customer focused leasing opportunities, driven by strategic change in the business environment identified in Chapter 1, and the volatility of the business environment. Businesses need to be more flexible and agile with the ability to 'pay as you go' in some circumstances for space and business services.

For major organizations the planning of procurement is both critical and complex. In a modern flexible organization that is project-based the need for space may be highly volatile both in terms of space requirements and location. Strategic planning of facilities may need to adopt a 'hub and satellite' or 'core and flex' approach with core functions and headquarters anchored in strategic locations in longer-lease or freehold buildings with greater flexibility around satellite operations, sub-offices and networks. Measuring space utilization, which we will discuss in Chapter 5, frequently reveals significant wastage in meeting rooms and ancillary space. We suggest that where possible meeting rooms should be procured on a similar basis with essential meeting rooms only being owned or leased and volatility being managed through flexible arrangements with landlords or operators such as Regus or MWB. As we explore in Case Study 1, contemporary landlords like Bruntwood offer clients meeting facilities on a 'pay as you go' basis rather than having to include them within their leased premises.

Land Securities Landflex

Landflex was a contemporary leasing solution launched by Land Securities in 2003; it is now available on five office properties in London:

47 Mark Lane, EC3
7 Soho Square, W1
10 Eastbourne Terrace, W2
40 Eastbourne Terrace, W2
140 Aldersgate Street, EC1

Landflex may be a model for the future as occupiers demand greater flexibility, cost certainty and transparency in their leasing arrangements. According to Landflex, they offer an emphasis on Customer Service and Building Management, aiming 'to take the headache out of running your business accommodation'.

Landflex was designed to offer leases tailored to business needs and can even provide a *'blend of leases'* within a single building allowing for core and flex operations and/or project teams. The way in which Landflex differs from traditional leasing is set out in Table 3.5.

Landflex customers pay a single monthly accommodation charge, which includes:

- Rent
- Insurance
- Core services
- Reception
- Cleaning of demise
- Category 6 cabling
- Maintenance of building and demise
- Post room
- Security
- Utilities
- Carpets

TABLE 3.5 The Differences between Traditional Leases and Landflex Leases

Traditional lease management	Landflex
Usually only one lease is on offer covering all the space you wish to take from day 1	Many leases may be taken at the same time in the same building
Lease length is typically 10 years, but can be 15 or 5 years	Leases may be of any length between 1 year and 15 years
Leases may contain one break, typically half way through the term. Renewals are a right under the Landlord and Tenant Act	Leases may contain options to cancel (break) or extend (renew)
Repairs and maintenance are the responsibility of the tenant, which will usually result in dilapidations at the end of the term	Repairs and maintenance are the responsibility of Landflex, meaning no dilapidations bills at the end of the term
Rent is reviewed at the end of the fifth year to the then prevailing open market rent. Service charges vary annually	Rent and core services are reviewed annually upwards and downwards in line with the RPI
Assignment may result in continued liability until the end of the term	Assignment achieved by surrender and re-grant

The core services provide a suitable modern working environment for organizations and their employees and cover services to both the common parts and customer offices, including:

- Landflex building management
- Customer office and Landflex building specification
- Maintenance and life cycle replacement to building and customer offices
- Cleaning to building and customer offices
- Building security
- Building reception and visitor management
- Building insurance
- Utilities – small power and lighting
- Mail and post room services

and the option of to procure a number of additional services as required.

- Fit out
- Furniture
- Business Club
- IT (Voice/Data)
- Catering

Effectively the only additional costs to occupation are business rates (UK property taxes).

Landflex also claim to offer both price certainty and a lower cost than traditional leasing. Landflex aim to ensure that their customers know from the outset what their office occupancy costs will be for the duration of their occupation. The only uncertainty being what the retail price index will be which is used to index the rent throughout the occupation.

Landflex produce price comparisons verified by the IPD Occupiers Property Databank (see Chapter 5) which suggest that the Landflex product saves 10% of occupancy costs at their 7 Soho Square building, compared to a similar building on a traditional lease. This will partly be attributable to the provision of flexible meeting room space which does not have to be leased by the occupier. These services are provided by the Landflex Business Club and can be paid for on an hourly or daily basis.

However, there are some disadvantages. One limitation to the leases is that no structural alteration can be made to the occupier's demise. Consent from Landflex for alterations has proved to be very difficult to obtain, so tailoring and branding may be problematic. In addition, Landflex is currently only available in London.

Regus and MWB virtual offices

Both Regus and MWB and Bruntwood are now providing 'virtual office' solutions to complement their highly flexible office products. Regus Virtual Office claims to 'provide your business with all the benefits of a prime office, without the need to be there'.

Companies can now use one or more of the Regus network of prestigious addresses worldwide, ensuring the best impression for businesses and providing professional front line support. Regus Virtual Office (Plus) provides companies with a part-time office solution, designed to maximize time and working effectiveness, at a fraction of the cost of having a permanent office.

Regus or MWB manage telephone calls, faxes, messages and mail, and when companies need a prestigious location for a meeting, they have access to offices, meeting rooms, video conferencing facilities on a 'pay-as-you-use' basis. In the case of Regus, customers have access to more than 900 business centres across the world.

With modern global business demanding that more and more employees spend time away from the office these solutions may be appropriate for highly mobile organizations.

Regus Business World

Regus has broken further new ground through the launch of its Business World membership programme in June 2008, allowing members to take full advantage of this unrivalled network of global offices, one of Regus's fundamental strengths. Regus has always believed that the future lies in responding to changing needs, particularly the growing demand for 'flexibility and dynamic solutions'. Business

World membership allows companies to utilize the benefits and facilities provided by Regus and offers access to over 950 business lounges and private offices across the world. The membership type dictates the level of discount and access to facilities. The membership options are set out in Table 3.6.

Converting from Freehold to Leasehold

SALE-AND-LEASEBACK

The basic sale-and-leaseback transaction is defined by Adams & Clarke (1996) as one in which:

> *the owner of a property sells that property to a third party and simultaneously takes a lease on that property from the third party.*

Reported to be first used in the UK in the late 1920s and with the biggest large-scale application in the USA by Safeway in 1936 (Adams & Clarke 1996), sale-and-leaseback has become a staple strategic option for real estate. In more recent times, a sample of transactions reported in *Estates Gazette Interactive* demonstrates the variety of sale-and-leaseback transactions:

- September 2004: Thirty-eight off-licences from the Thresher Group for sale at auction at Allsops on a sale-and-leaseback basis.

TABLE 3.6 Regus Business World Membership Types, Benefits and Costs

Membership type	Benefits	Cost (as at 2009)
Blue	5% discount on meeting rooms, day rooms and video conferencing facilities	free
Gold	Unlimited walk-in access to 950 business lounges and cafes worldwide with complimentary Internet and refreshments 10% discount on meeting rooms, day rooms and video conferencing facilities Access to professional, administrative support from packing and shipping to document printing and binding	£199 per year
Platinum	10 days of access to a private office in any of Regus's 950 business centres worldwide Plus all the benefits of Gold membership	£299 per month £2999 per annum
Platinum Plus	Unlimited access to a private office in any of Regus's 950 business centres worldwide Plus all the benefits of Gold membership	£499 per month £4999 per annum

- October 2004: IBIS Hotels £250 million sale-and-leaseback to London and Regional.
- March 2005: Debenhams £495 million sale-and-leaseback to British Land.
- August 2005: Boots £298 million sale-and-leaseback to Reit Asset Management.
  - November 2005: Tesco £366 million sale-and-leaseback to Morley Fund Management.
  - December 2007: Italian textiles group Miro Radici Group €55.5 million sale-and-leaseback of logistics warehouses in Germany to Sellar Property Group.
  - December 2008: Tesco £109 million sale-and-leaseback to PRUPIM annuity fund.
  - October 2009: HSBC $330 million sale of New York HQ to special purpose vehicle 452 Fifth Avenue in a sale-and-leaseback transaction to Midtown Equities and Israeli tycoon Nochi Dankner.
  - November 2009: Unite Group £21.5 million sale of student accommodation to M&G Secured Property Income Fund.

These transactions demonstrate that there is a continuous appetite for the right sale-and-leaseback, and as KPMG (2006) comment: 'sale and leaseback was once perceived as the borrowing of last resort. It has now grown in status … with the recognition of the often significant funding potential within assets whose value has often appreciated markedly over recent years.'

KPMG (2006) identify the following potential benefits for the seller from an accountancy perspective:

- frees up capital to fund merger and acquisition activity or to invest in core business activities;
- tax benefits are realized by offsetting lease costs as an operating expense;
- seller remains in day-to-day operational control of the property;
- improvement in balance sheet through exchange of fixed assets, often carried out at a below market value, for cash;
- one-off Profit and Loss account benefit reflecting profit realized over book value;
- transfers property value risk to a third party on a fully transparent basis.

We identify a number of potential disadvantages:

- the potential loss of capital allowances;
- partial loss of control of the assets;
- loss of any capital growth potential;
- exposure to rental growth and upward only rent reviews (in the UK);
- loss of an asset that can be used to secure further borrowing;
- uncertainty at the end of the lease term.

The Impact upon Company Accounts

The focus of much of the literature on sale-and-leasebacks is around the generation of capital. In addition, the removal of real estate from the Fixed Assets of the Balance Sheet will have a significant impact upon the financial statements of a company. This is a complex area, which we examine here using a simplified scenario to illustrate the main accounting impacts.

Simplified Sale-and-Leaseback Scenario

A company is considering selling and leasing back their headquarters, which has been valued at £1,000,000.

It has agreed to a sale-and-leaseback on a 10-year lease at £100,000 p.a.

The operating profit of the company is £600,000.

The situation **BEFORE** the sale-and-leaseback in the Balance Sheet is shown below:

Fixed Assets	£2,000,000 (including the £1,000,000 office building)
Current Assets	£500,000
Total Assets	**£2,500,000**
Current Liabilities	£600,000
Long Term Liabilities	£200,000
Net Assets	**£1,700,000**

Calculating two of the key accounting ratios indicates:

ROCE = 31.5%	**Operating Profit/(Total Assets − Current Liabilities) × 100** **(600,000/1,900,000) × 100 = 31.5%**
GEARING = 11.76%	**Long Term Liabilities/Net Assets × 100** **(200,000/1,700,000) × 100 = 11.76%**

Undertaking a sale-and-leaseback will in simple terms remove the £1,000,000 from the Fixed Assets. The capital may be used to pay off a mortgage which will reduce the impact on gearing, which is not shown in this example; or the capital receipt may be accounted for by special provisions. A rental liability will be added to the current liabilities.

Current accounting standards provide for treatment of leases in two different ways under International Accounting Standard (IAS) 17. If an operating lease is created then the lease remains off the balance sheet and this is the usual situation for a real estate lease. However, if a finance lease is created, defined as where the lease transfers substantially all the risks and rewards incident to ownership, then the lease will be recapitalized in the balance sheet using methods described in IAS17.

The situation **AFTER** the sale-and-leaseback in the Balance Sheet is shown below (assuming an operating lease is created):

Fixed Assets £1,000,000	(£1,000,000 removed)
Current Assets	£500,000
Total Assets	**£1,500,000**
Current Liabilities	£700,000 (£100,000 rental added)
Long Term Liabilities	£200,000
Net Assets	**£600,000**

Recalculating two of the key accounting ratios indicates:

ROCE = 75%	**Operating Profit/(Total Assets − Current Liabilities) × 100** (600,000/800,000) × 100 = 75%
GEARING = 33%	**Long Term Liabilities/Net Assets × 100** (200,000/600,000) × 100 = 33.33%

NOTE: The above example assumes an operating lease will be used and is in accordance with accounting standards at the time of publication. Later in this chapter we examine the significant changes being proposed to the treatment of operating leases in company accounts by the International Accounting Standards Board.

Contemporary Example of a Sale-and-Leaseback Transaction Designed to Engineer a 'Win–Win' Outcome for All Stakeholders

In 2009 the Gucci Group N.V. instructed Cushman & Wakefield to introduce their freehold premises at 32/33 Old Bond Street, London W1 to the investment market.

The property comprised:

- a flagship retail store of 678 square metres arranged over basement, ground and first floors occupied by Yves Saint Laurent (part of the Gucci Group); and
- offices on the third to sixth floors, totalling 1390 square metres, also occupied by the Gucci group.

The transaction was designed to create a sale-and-leaseback arrangement on the following new leases on completion of the sale to the investor:

- Retail component: a 15-year lease on full repairing and insuring terms, with the benefit of five yearly upward only reviews.
- Office component: a 15-year lease on full repairing and insuring terms, with the benefit of five yearly upward only reviews.

The retail component was subject to a number of minor sub-tenancies.

The interesting component of this transaction was an initial low rent, designed to cushion the introduction of the lease costs upon the company accounts with a guaranteed uplift, secure income and a company guarantee from the parent company. Interestingly the investment brochure included the Dun & Bradstreet rating of the Gucci Group N.V. at 5A1, the highest rating achievable, reflecting the minimal risk of business failure.

Advantages for the Gucci Group in conducting this Sale and Leaseback transaction:

- release of capital: offers in excess of £26.5 million;
- introduction of an initial low rent during a challenging trading period for retailers with uplift anticipated to coincide with a return in confidence;
- cushioning of the introduction of rental expenses in the company accounts.

Whilst a number of offers were received, some reported to be in excess of the £26.5 million required, the Gucci group did not proceed with the transaction. Although it is not known why, one may speculate that uncertainties over the proposed changes to the accounting of leases on the balance sheet may have had an influence on the decision.

SALE-AND-MANAGEBACK

Sale-and-manageback has become popular in the leisure industry, particularly hotels where it offers additional advantages. The sale-and-manageback is defined by Tipping & Bullard (2007):

> *This model involves the operating company selling its property to an investor, which then grants a management contract, instead of a subsidiary interest of the property, to the operator.*

Billions of pounds worth of hotels have been placed into management structures by the large hotel chains in recent years (Schäfer-Surén, 2005). This model, currently avoids leases appearing as liabilities on the balance sheet and avoidance of certain taxes including Stamp Duty Land Tax in the UK. Hilton Hotels completed a £400 million manageback of 16 UK hotels in 2005.

The sale-and-manageback is more sophisticated than the traditional sale-and-leaseback in that it frequently grants the investor a share of the turnover instead of, or in addition to, rental income.

Future Implications for Procurement Options

The sale-and-leaseback section has shown how the movement of real estate off the balance sheet not only provides access to capital that a company may use more effectively in its business but can also improve the way in which the company is assessed in terms of its balance sheet and the key business indicators that are applied to the company accounts.

However, organizations should be aware of the impending global changes being proposed by the International Standards Board (IASB) and the Financial Accounting Standards Board (FASB) in the United States. The following information and example is based upon information very kindly provided by Richard Porter, Associate Director, Corporate Capital Markets, Jones Lang LaSalle, London and presented at a CoreNet Global Seminar in London, October 2009 by Richard and Rachel Knubley, Senior Project Managers, IASB.

IASB and FASB are working towards a policy where the current distinction between operating and finance leases is abolished. Currently operating leases have no provision in the balance sheet and this has resulted in sustained interest in sale-and-leaseback arrangements and a shift away from freehold ownership (Deloitte, 2009). The legislation aims to create greater economic transparency and consistency in financial statements.

Basically the new standards will provide a single method of accounting for leases and the current operating lease model will disappear. All leases will follow a 'right of use' model in which the lessee is assumed to purchase a right to use the leased item and which recognizes:

- a right of use asset;
- an obligation to pay rentals.

There remain some uncertainties in terms of the detail of the proposals and further consultation will take place. However, the basic components are unlikely to change and they have considerable ramifications for real estate managers and in particular have a significant impact upon surrender and renewal transactions. The main impact for real estate is the removal of the way in which operating leases are accounted for. In future all leases will be accounted for in the same way. These changes are summarized in Table 3.7.

Using a simplified model adapted from one kindly provided by Richard Porter of Jones Lang LaSalle we can examine the impact of the changes to a single asset upon the Profit and Loss, balance sheet and cash flow accounts.

The example is based upon a property with:

- a 10-year lease;
- an initial open market rent of £1,000,000 per annum;
- an incremental borrowing rate of 5%.

The BEFORE accounts are straightforward and are set out in Figure 3.3. The asset remains off the balance sheet, other than the cumulative cash relating to the rental payments.

Turning to the situation after the proposed IASB changes are implemented requires consideration of a number of additional calculations that have to be completed.

TABLE 3.7 Proposed Changes in the Way Leases Will Be Accounted for in Company Accounts

Operating leases *today*	Operating leases *tomorrow*
Rent is expressed in the Profit and Loss account	An asset appears on the balance sheet as in ownership. It is the 'right of use' asset
The obligation to pay rent is not recorded in the balance sheet	The asset is depreciated over the term (straight line)
Underlying asset not recorded on balance sheet	A liability currently referred to as 'obligations to pay rentals' appears on the balance sheet – treated like a loan
Obligations usually reflected in the notes to the accounts	Liability is repaid over the term, through rental payments
	Payments are split into Interest and Capital. Interest is a P&L expense, capital reduces the liability P&L cost is Interest and depreciation of the asset

Operating Lease – Example in accordance with current regulations

PROFIT and LOSS		Year 0	Year 1	Year 2		Year 10
	Rent		(1,000,000)	(1,000,000)		(1,000,000)
	P&L		(1,000,000)	(1,000,000)		(1,000,000)
CASH FLOW		Year 0	Year 1	Year 2		Year 10
	Rent		(1,000,000)	(1,000,000)		(1,000,000)
	CASH FLOW		(1,000,000)	(1,000,000)		(1,000,000)
BALANCE SHEET		Year 0	Year 1	Year 2		Year 10
	Cumulative Cash		(1,000,000)	(1,000,000)		(1,000,000)
	BALANCE SHEET		(1,000,000)	(2,000,000)		(10,000,000)

Figure 3.3 Accounts BEFORE the proposed changes are applied

To calculate the capitalization of the asset, or the *'right of use'*, and the liabilities or *'obligations to pay rentals'*, the liabilities are discounted over the lease period at the incremental borrowing rate:

This discounted cash flow calculation calculates the present value of the asset in terms of the rental outgoings, discounted at the incremental borrowing rate.

The NPV of £7,721,735 is used as the *'right of use'* value in the Balance Sheet.

Year	Rental	Present Value @ 5%	Discounted Outgoings
1	1,000,000	0.95	952,381
2	1,000,000	0.91	907,029
3	1,000,000	0.86	863,838
4	1,000,000	0.82	822,702
5	1,000,000	0.78	783,526
6	1,000,000	0.75	746,215
7	1,000,000	0.71	710,681
8	1,000,000	0.68	676,839
9	1,000,000	0.64	644,609
10	1,000,000	0.61	613,913
		NPV=	**7,721,735**

To calculate the obligations to pay rentals the above calculation must be adjusted for each accounting year. So for the next period (Year 1) when there are nine rental payments outstanding the calculation would be:

Year	Rental	Present Value @ 5%	Discounted Outgoings
1	1,000,000	0.95	952,381
2	1,000,000	0.91	907,029
3	1,000,000	0.86	863,838
4	1,000,000	0.82	822,702
5	1,000,000	0.78	783,526
6	1,000,000	0.75	746,215
7	1,000,000	0.71	710,681
8	1,000,000	0.68	676,839
9	1,000,000	0.64	644,609
		NPV=	**7,107,822**

The new NPV represents the obligation to pay rentals for Year 1; this process is repeated for each year of the lease.

In Year 2 the figure of £7,107,822 is inserted into the Balance Sheet as the obligation to pay rental liability.

In Year 3 with eight periods of rent, recalculation produces an obligation to pay a rental figure of £6,463,213.

Having calculated the net present value this figure can be used to calculate straight line depreciation which must be included in the Profit and Loss account. Dividing the above present value by ten results in an annual depreciation of £772,173.

Next the notional interest expense has to be calculated. This is calculated as if the present value amount was being financed on a repayment mortgage. Only the interest payments are included in the Profit and Loss account as an 'interest expense'. This has been simplified below on an annual basis of capital adjustment.

Year		Capital repaid	Interest @ 5%
1	7,721,735		386,087
2	6,863,735	858000	343,187
3	6,005,735	858000	300,287
4	5,147,735	858000	257,387
5	4,289,735	858000	214,487
6	3,431,735	858000	171,587
7	2,573,735	858000	128,687
8	1,715,735	858000	85,787
9	857,735	858000	42,887
10	−265	858000	−13

The accounts can now be presented using the above calculations and in accordance with the proposed IASB changes and are set out in Figure 3.4:

Accounts under new provisions:

PROFIT and LOSS		Year 0	Year 1	Year 2		Year 10
	Depreciation		(772,173)	(772,173)		(772,173)
	Interest Expense		(386,087)	(343,187)		(0)
	P & L	-	(1,158,260)	(1,115,360)		(772,173)
Note that the Profit and Loss is different from the current methodology. It is initially higher but falls over time. Under the current regulations the P&L is constant at £1,000,000.						
CASH FLOW		Year 0	Year 1	Year 2		Year 10
	Terminology to be confirmed		(1,000,000)	(1,000,000)		(1,000,000)
	CASH FLOW		(1,000,000)	(1,000,000)		(1,000,000)
Note that the Cash Flow is no different from the current methodology.						
BALANCE SHEET		Year 0	Year 1	Year 2		Year 10
FIXED ASSETS	'Right of Use' asset initial capitalization	7,721,735	7,721,735	7,721,735		7,721,735
	Cumulative depreciation	-	(772,173)	(1,544,346)		(7,721,735)
Net Book Value		**7,721,735**	**6,949,562**	**6,177,389**		**0**
CASH	Cumulative cash flow	-	(1,000,000)	(2,000,000)		(10,000,000)
LIABILITIES	Obligation to pay rentals	(7,721,735)	(7,107,822)	(6,463,213)		0.0
	BALANCE SHEET	0.0	(1,158,260)	(2,285,824)		(10,000,000)
Note that the balance sheet is significantly different from the current methodology.						

Figure 3.4 Accounts AFTER the proposed changes are applied

The Commercial Implications of the Changes

As can be seen by the comparison of the two sets of accounts, there are significant differences in both the Profit and Loss and the balance sheet. There are therefore a number of significant implications for corporate real estate asset managers when the IASB proposals are introduced, including:

- assets sold under sale-and-leaseback arrangements will come back onto the balance sheet, potentially making such transactions less attractive;
- the Profit and Loss account will suffer an initial shock which will be more dramatic for longer leases;
- the balance sheet will 'balloon' due to the introduction of the right of use asset and corresponding liabilities, shorter leases will result in less ballooning;
- there may be a shift to increased ownership;
- occupiers will have to make careful decisions about lease options, breaks etc. as they have significant impacts upon the way the leases are treated;
- a debt load will be incurred for many companies;
- there will be an adverse impact upon key ratios such as ROCE and gearing.

To explore this final point we set out below a restatement of the highly simplified balance sheet for the sale-and-leaseback scenario examined earlier with the IASB changes applied and in Table 3.8 the ROCE and Gearing Ratios for the three situations examined.

TABLE 3.8 Examination of the ROCE and Gearing Ratios

Situation	ROCE	Gearing (assuming no mortgages or loans to be repaid from proceeds)
BEFORE Sale-and-leaseback	31.5%	11.76%
AFTER Sale-and-leaseback (with current operating lease provisions)	75%	33%
AFTER Sale-and-leaseback (with application of new IASB provisions)	35.9%	139%

Balance Sheet for Sale-and-Leaseback Scenario

Fixed Assets	£1,772,173	(£772,173 Right of Use asset valuation added back)
Current Assets	£500,000	
Total Assets	**£2,272,173**	
Current Liabilities	£600,000	
Long Term Liabilities	£972,173	(£772,173 obligations to pay rentals added)
Net Assets	**£700,000**	

Recalculating the two key accounting ratios indicates:

ROCE = 35.9%	Operating Profit/(Total Assets − Current Liabilities) × 100 (600,000/1,672,173) × 100 = 35.9%
GEARING = 139%	Long Term Liabilities/Net Assets × 100 (972,173/700,000) × 100 = 138.9%

We must stress that these examples are highly simplified and illustrative only. They do, however, provide an illustration of the accounting issues for sale-and-leaseback transactions and how the proposed IASB changes will significantly affect the company accounts of companies. It shows that companies who have been involved in significant sale-and-leaseback transactions, such as the major supermarkets, have some significant challenges ahead both in terms of the complexity and revaluations required and the initial impact upon their accounts in terms of both the Profit and Loss and the balance sheet.

Beyond Sale-and-Leaseback: Strategic Alternatives to Traditional Freehold or Leasehold Procurement

Total Property Outsourcing

Property outsourcing has been driven by the argument that real estate and its management is not the core business of an organization and is something that can be outsourced to a professional operator and converted into a more manageable cost at agreed levels of service delivery from the outsourcing company.

Probably one of the most significant early examples of outsourcing was the sale, in 2000, of all 1300 properties of Abbey National to Mapeley, a Guernsey-based property company specializing in property investment and outsourcing. In exchange Abbey received a 20-year package of leasebacks and management of leasehold property and the ability to immediately vacate 212 properties.

According to Mapeley (2009), outsourcing is defined as:

Outsourcing involves the transfer of property and the risks associated with occupying, managing and servicing that property, in return for occupational flexibility and straightforward charging.

The Abbey National Mapeley transaction was much more than a sale-and-leaseback to raise cash. Abbey needed a much more flexible format of occupation to manage an uncertain future as the financial services sector changed, particularly the introduction of Internet banking. It included property management services but not facilities management (FM) which it retained.

Louko (2004) suggests that organizations were generally successful in increasing their efficiency in space and capital use through outsourcing. However, it is reported that for all the firms studied in his research (including Abbey National) there was significant difficulty in predicting the future needs for flexibility. For example, in the case of Abbey National it was anticipated in 2000 that Internet banking would grow extremely quickly, therefore reducing the importance of a bank branch network. Internet banking was much slower to take off than anticipated and Abbey has been forced to stay in occupation far more than planned having paid a significant premium for flexibility.

Corporate PFI

The term corporate PFI emerged with the introduction of more sophisticated models of total property outsourcing by operators such as Land Securities Trillium. The BBC opted for a 30-year corporate PFI-style deal with LS Trillium, including both estate and FM of its 4 million square feet portfolio.

It followed the model of Private Finance Initiatives and Public–Private partnerships that were being used to procure hospitals and other public service operations by leveraging private investment and leasing back space to occupiers on a single occupancy charge basis covering all aspects of occupation; it usually involves the transfer of both real estate and FM operations. These transactions were of a very significant scale and complex in that the partnership did not just convert freehold to leasehold but often involved a complex strategy of development of underutilized assets, and the ability, in the case of the BBC, to generate major new facilities through the development of land at White City. Significantly it is the ability of the operator to raise cash in the capital markets to finance both the acquisition of the organization's real estate assets and its development that is the crucial component of this type of procurement. In the case of the early deals it is clear that Land Securities Trillium were probably as interested in the development opportunities within the BBC estate as they were in providing real estate and facilities services.

The reasons for corporate organizations adopting this approach are:

- an immediate release of capital, available for business growth and investment;
- increased efficiency, focusing on the core business;
- greater flexibility of occupation and opportunities for greater space utilization;
- increased profits and returns to shareholders;
- greater predictability of future real estate costs;
- reduction of annual occupancy and property operating costs;
- ability to downsize and improve facilities through a partnership which allows the new owner to exploit development opportunities.

There are some potential disadvantages, including:

- the cost of flexibility; in the case of Abbey National the initial rent roll represented an expensive 17.5% of the freehold value;
- if early emphasis was on reduction of costs, corporations may find themselves locked into significantly longer contracts (ABB & Swisscom); reducing flexibility of core operations;
- leases of unsuitable lengths;
- lack of control to refurbish or make alterations.

Norwich Union

Norwich Union part of the Aviva Group, the UK's largest insurer, also entered into a partnership with LS Trillium. A portfolio of 111,500 square metres of office accommodation throughout the UK, with an investment value in excess of £150 million was the subject of the deal with Norwich Union seeking a partner to assume responsibility for managing and improving their estate. The 25-year partnership involved:

- the transfer of ownership, management and improvement of 19 of Norwich Union's operational properties to Trillium;
- comprehensive refurbishment of strategic parts of the portfolio;
- transfer of real estate risks;
- all future maintenance and capital works transferred with greater costs certainty;
- management resources released to focus on their core business;
- revolutionary revitalization of Norwich Union's head office creating a modern, environmentally sustainable hub in the heart of Norwich.

Benefits to Norwich Union

- improved, more productive working environment;
- greater cost certainty;
- focus on core business;
- release of capital and unlocking of development gain.

Stuart Smith, former Director of Occupied Property Strategy, Norwich Union, summed it up by stating: 'Trillium is providing a better working environment for our staff. This deal allows Norwich Union to focus on our core business of delivering insurance services to our customers.'

The Corporate PFI movement has not been without problems and controversy; it is safe to say that the jury is still out. Few transactions have taken place since 2007 and significantly Land Securities has sold its Trillium business. This suggests that this model relies on a buoyant property market for the operators to be able to extract enough development and latent value in operational portfolios for the deal to be sufficiently attractive. In fact the BBC terminated their £2.5 billion deal with LS Trillium in May 2005. Tara Conlan in the *Media Guardian* commented:

> *Just three-and-a-half months into the deal, BBC chief operating officer John Smith has confirmed that the contract is ending. The BBC no longer needed Land Securities to help it raise money in the bond market for its building works. Mr Smith said 'Some things have changed since 2001. The main thing is we can get access to the bond market now. We have an AA rating and can get a cheaper rate than Land Securities. That set us thinking we did not need to outsource, so we started rethinking the agreement'.*

There were overspends and contractual problems with some of the redevelopment works which may have added to the reasons for cancelling the contract. LS Trillium did make a £23 million profit through selling back the White City media village to the BBC. The BBC subsequently put out to tender a £900 million, 10-year real estate management contract to replace the LS Trillium deal.

Summary Checklist

1. Organizations should continually monitor their balance of property procurement, especially the leasehold to freehold ratio. How does your organization compare to the Lasfer Curve and its 65% leasehold to 35% freehold optimum distribution?
2. Examine the advantages and disadvantages of Freehold and Leasehold property and the strategic and financial issues; how does this relate to your organization, or its business units. Are you in the right type of ownership?

3. Do you have sufficient flexibility within your portfolio? Would a more contemporary approach be beneficial?
4. Are the lengths of the leases of your operational real estate consistent with your business plans? Do you have sufficient flexibility and have you considered the real risks of long leases compared to short options
5. Undertake utilization studies of your meeting and boardrooms. Are they being underutilized? If so you should consider moving to premises, at the next opportunity, where you can procure these and other ancillary space on a 'pay as you go' basis.
6. If you are considering a new building examine the cash flows of both freehold and leasehold opportunities in addition to the strategic and operational needs.
7. If you consider that capital tied up in your portfolio is not working as hard as capital in the business it may be worthwhile examining a sale-and-leaseback transaction.
8. If you do consider sale-and-leaseback consider the implications of the impending IASB changes to accounting for leases in company accounts and consider how this may affect the attractiveness of a sale-and-leaseback.
9. Large organizations may wish to consider total property outsourcing and focus on their core business. However, should be aware of both the advantages and disadvantages to such transactions.

References

Adams, A., & Clarke, R. (1996). Stock market reaction to sale and leaseback announcements in the UK. *Journal of Property Research, 13*, 31–46.

Collins, D., & Montgomery, C. (1995). Competing on resources. *Harvard Business Review*, 108–128, July–August.

Conlan, T. (2005). *BBC cancels Land Securities contract*. www.guardian.co.uk/media/2005/may/12/broadcasting.bbc3. Accessed 12.11.09.

Cooke, H., & Woodhead, S. (2007). *Corporate occupiers handbook*. London: EG Books.

Deloitte. (2009). *Summaries of international financial reporting standards (IAS17)*. www.iasplus.com/standards/ias17.htm. Accessed 12.11.09.

Devaney, S., & Lizieri, C. (2004). Sale and leaseback, asset outsourcing and capital markets impacts. *Journal of Corporate Real Estate, 6*(2), 118–132.

Ellis, I. (2004). *Land Securities Trillium and Norwich Union agree new property outsourcing deal*. http://www.landsecurities.com/press. Accessed 20.11.09.

Estates Gazette Interactive (Subscription based information service) www.egi.co.uk [online]. Numerous searches of sale-and-leaseback transactions, retrieved October and November 2009.

IDRC Foundation, *Managing for Shareholder Value*: A Summary of Proceedings from IDRC's Executive, Research Bulletin No. 18 in the series by The IDRC Foundation. (ISBN): 0-931566-60-6.

KPMG. (2006). *Corporate finance: sale and leaseback takes the spotlight*. www.kpmg.co.uk/pubs/sale and leaseback.pdf. Accessed 05.10.09.

Lasfer, M. (2007). On the financial drivers and implications of leasing real estate assets. The Donaldsons–Lasfer's Curve. *Journal of Corporate Real Estate, 9*(2), 72–96.

Linneman, P. (1998). Corporate real estate strategies. Real Estate Department, The Wharton School, University of Pennsylvania. http://knowledge.wharton.upenn.edu/papers/442.pdf. Accessed 05.07.09.

Louko, A. (2004). Competitive advantage from corporate real estate disposals. *International Journal of Strategic Property Management, 8*(1), 11–24.

Mapeley. (2009). *Outsourcing*. www.mapeley.com/property/outsourcing/default.aspx. Accessed 05.07.09.

Porter, R. (2009). *Jones LangLaSalle/CoreNet Global Seminar. Perspectives on operating leases changes.* Attended October 2009, retrieved from www.corenetglbal.org.uk 05.11.09.

Schäfer-Surén, J. (2005). Make the right choice from hotel lease menu. *Estates Gazette, 548*, 48.

Tipping, M., & Bullard, R. (2007). Sale and Leaseback as a British real estate model. *Journal of Corporate Real Estate, 9*(4), 205–217.

Weatherhead, M. (1997). *Real estate in corporate strategy.* Basingstoke: Palgrave.

Chapter 4

Customer Focused Procurement Options

Introduction

Customer focused procurement options were briefly introduced in Chapter 3 in relation to innovative landlords such as Landflex, MWB and Regus. We believe that customer focused landlords offer significant strategic advantage to their customers and this is explored in this chapter and in Case Study 1, which examines the Bruntwood approach.

In this chapter we aim to explore the advantages of choosing a landlord that has a more customer focused approach, the benefits that this can bring and strategies for maximizing these benefits. We start by reviewing the innovators other than Bruntwood who have helped shape the thinking on this approach, including Goodman (formerly known as Arlington), BAA and, more recently, Fasset property management. We can see a strong connection between the development of customer focused lease delivery and the drivers of change introduced in Chapter 1.

Changing Expectations

The world of landlord and tenant relationships has changed dramatically over the past 30 years. This is driven during recessionary times out of necessity but also by the changing expectations of tenants; the resistance of occupiers to traditional long institutional leases in a global economy; and businesses operating on shorter and shorter business planning cycles, needing more flexibility. In the UK, in particular, the standard 25 years' institutional lease, so beloved by investors, has become a barrier to business and an area of concern for the UK government, which perceives it as an obstacle to inward investment and the creation of an enterprise culture. A number of measures, as yet voluntary, such as the Commercial Lease Code, have been introduced to make UK leases more attractive and consistent with global expectations.

Corporate Real Estate Asset Management. ISBN: 978-0-7282-0573-4

The Unique UK Perspective and Why It Is Changing

The typical UK landlord has for many years had a very different approach from many other parts of the world. This is partly because of the cultural heritage of the UK feudal system; and the development of a sophisticated investment property market. This market demanded high levels of income security through long leases with very little active management input, other than the collection of rent, until the end of the lease. We feel compelled to start our discussion with the delightful quote attributed to the late Howard Bibby, then the managing director of Arlington Business Services, a major business space developer and manager in the UK. He was interviewed by *Premises & Facilities Management* magazine in January 1999. His comments and criticisms of UK landlords set the scene for this chapter. Unfortunately they remain true for some landlords, although the recession of 2007 onwards is probably precipitating more change in a customer focused direction.

Howard Bibby (1999) believed that we were then beginning to see an unstoppable 'convergence of property and FM, with a new focus on the customer'. He believed that landlords could no longer ignore tenants as customers. Using some pretty strong and significant terms he went on to say:

> *The average corporate doesn't feel well served by the property industry. It is seen as inflexible, bureaucratic, shambolic, reactive and reactionary. With few exceptions, tenants are still treated as revolting peasants. Most development companies, particularly trader-developers, see themselves as providing investment product, the tenant is incidental.*

Bibby, was a visionary, believing that occupiers are getting more sophisticated and demanding, asking key questions such as: 'If I take this building, can I retain and attract my key staff?' This question remains central to our approach to corporate real estate; occupiers do have a choice between traditional passive landlords and those focused on the customer, such as Bruntwood in Manchester, the subject of our Case Study 1.

This choice is being explored with reference to staff motivation, staff retention and issues of productivity all of which is explored in this book and has a direct link to the philosophy and delivery mechanisms of contemporary landlords. Bibby (1999) was also very astute in predicting that

> *changing funding arrangements based on unitization and Real Estate Investment Trusts will unite investment with a service culture forcing developers to take the future management of buildings more seriously.*

Satisfaction with UK Landlords

Satisfaction with UK leases and lease management practice has for many years been low. In 2001 a survey by the RICS (Crosby, Gibson & Oughton, 2001) revealed that only 8% of respondents thought the UK leasing system to be satisfactory.

The Property Industry Alliance and Corenet Global UK, through Real Service and IPD, have for the past 3 years introduced a UK Occupier Satisfaction Index (OSI). Through interviews with tenants it explores a number of issues and creates an on-going annual satisfaction score. This satisfaction score is defined as:

The ability of the supply side of the UK commercial property industry to deliver the products and services that its occupier customers require.

The OSI Index overall scores published so far are set out in Table 4.1:

The figures in Table 4.2 show that overall satisfaction is mainly below 60% and is remaining relatively static. Within the detail of the study the following table

TABLE 4.1 Overall Satisfaction Scores, 2007–2009

OSI	2007	2008	2009
All occupiers	55	57	57
Industrial	55	54	54
Office	59	61	62
Retail	52	57	55

TABLE 4.2 Extracts from the 2009 Occupiers Satisfaction Survey

Topic	% change from 2008	Respondents' comments
Understanding of occupier needs	+15% change in 2009	Occupiers would like more contact with property owners, and for them to show more interest, empathy and understanding
Responsiveness	+7% change in 2009	Occupiers see some improvement but many have to chase to get things done
Overall value for money	+1% change in 2009	Value for money is the key concern and occupiers do not believe that the UK property industry is reacting fast enough to changes in the economic climate
Service charge value for money	Zero change in 2009	Occupiers want property suppliers to place greater emphasis on reducing costs and to consult with them more

indicates that the customer focused solutions identified in this chapter remain the exception rather than the rule. In Case Study 1 we examine the recommendation survey approach adopted by customer focused companies such as Bruntwood. It is estimated that the OSI scores would indicate a net recommendation score of −3%, i.e. more customers would not recommend their landlords to other customers than those who would.

The survey indicates that levels of dissatisfaction remain high in the areas shown in Table 4.3.

The results for 2009 help to create the agenda for this chapter and will also be used to examine our exemplar of customer focused management, Bruntwood, in Case Study 1. The dissatisfaction with service charge operations in the survey has compelled us to include this as well as compliance with the Commercial Lease Code as discrete elements of the current dissatisfaction with customer servicing.

Traditional *vs.* Contemporary Property Management

Having examined the lack of customer focus in traditional real estate management and, briefly, the drivers for a change of approach, we now aim to explain the contrasts between the traditional and the customer focused approach. A long standing definition of traditional property management can be found in Scarrett (1983):

> *Property management seeks to control property interests having regard to the purpose for which the interest is held: to negotiate lettings and to initiate and negotiate rent reviews and lease renewals, to oversee physical maintenance and enforcement of lease covenants, to be mindful of the necessity of upgrading and merging interests where possible, to recognize opportunities for the development of potential and to fulfil the owner's legal and social duties in the community.*

Notice the use of the word control and the focus on lease management transactions such as rent reviews, it is very pro landlord and does not promote the kind of win–win customer focused relationship we see in the contemporary approach. This

TABLE 4.3 Dissatisfaction Expressed in the 2009 Occupiers Satisfaction Survey

Topic	% Dissatisfied
Feel a valued customer	39%
Responsiveness to requests for service	36%
Value for money for service charges	50%

TABLE 4.4 Comparison Between the Landlord and Tenant Perspectives of Real Estate

Landlord perspective	Tenant perspective
Property viewed as income-producing asset	Property viewed as facility
Maximize income	Minimize occupancy costs
Certainty of income and cash flow	Certainty of occupation
Privity of contract (now replaced with guarantees to preserve liability to original tenants)	No contingent liabilities when lease expires or assigned to a third party
Avoid vacancies and voids	Flexible terms, break clauses and exit strategy

type of approach does nothing to reconcile the conflicts of a landlord and tenant relationship defined, for example by Edington (1997) and set out in Table 4.4.

We believe that to achieve workplace transformation and productivity as discussed in Chapter 6 working with a contemporary landlord, with a customer focus, may make it much easier to achieve your objectives.

The following comparison, set out in Table 4.5, examines the focus of traditional property management upon the asset and contemporary approaches which focus on the customer. We acknowledge work done by the late Howard Bibby (published as a case study on his website) and Edington (1997) in the case study of BAA as a basis for this table, which we have updated.

We believe that post 2010 as economies pull out of recession there is no place for the traditional approach to managing tenants. With shorter lease lengths and a business culture of flexibility and connectivity there needs to be a greater focus on cash flow rather than simple income security and this demands that the relationship should mature. The latest edition of the Occupiers Satisfaction Index validates this and points to a continuing 'can do better' response from UK tenants. We aim to set out in this chapter and illustrate with the Bruntwood Case Study why a customer focused approach leads to a win–win partnership with added value for both parties.

The Financial Argument to Customer Focused Management

Tenant retention became a major issue in the recession of the late eighties/early nineties and some organizations recognized this and created significant strategic advantage by doing so. The major recession from 2007 once again has highlighted the problems of losing tenants and, as our Case Study on Bruntwood confirms, those firms who have undertaken a customer focused approach have outperformed traditional landlords and better preserved their rental cash flow.

TABLE 4.5 A Comparison of Traditional and Customer Focused Real Estate Management

Traditional real estate management Focused on the asset	Customer focused real estate management Focused on the customer
The asset is managed to maximize income and capital returns and it is viewed primarily as a property investment	The property is managed as an envelope for the customer's business. Focus is on cash flow and tenant retention which is seen as the vehicle for stable investment returns and growth
The Landlord has a dismissive, distant and high handed approach and is ignorant of tenant's requirements or business needs. Tenants are regarded as a cash cow, no regard for building a relationship	Landlord sees the building of a positive relationship with its customers as pivotal in being successful. Regular positive communication underpins this approach
Limited communication only related to transactions and problems, such as non payment of rent.	Constant and pro-active interaction, dialogue and a win–win attitude to communications, there to solve problems for a mutually beneficial outcome
Anonymous buildings with no landlord brand identity	Branded buildings and retailing destinations with a sense of productivity and excitement
Adversarial, transactional relationship conducted at a distance with limited transparency	Open relationships, transparency, trusted, ethical, long term relationships
Periodic communication on operational or investment issues. Aggressive 5 year rent review negotiations	Continuous partnering dialogue to help maximize occupiers workspace performance and pay premium rents, supported by fully transparent 24/7 service desk and website
No service desk or formal complaints procedure. Service requests not managed to completion. Tenants kept in the dark	All service requests and complaints logged, managed and resolved in some cases through online service desk. Full transparency of progress
NO service provisions such as service level agreements, minimum standards or benchmarks	Service level agreements with benchmark targets and minimum specification. Penalties for non compliance (BAA)
On site staff (if present) engaged on operational and compliance issues. Staff recruited from traditional property backgrounds	On site staff focused on creating and managing sustainable relationships. Staff recruited from service backgrounds – e.g. hotel industry
Silos of hierarchical job functions containing fund manager, portfolio manager, asset manager, property manager, accounts and services manager	Integrated multidisciplinary teams with all participants having equal status

Financial Rationale for a Customer Focused Approach that Promotes Tenant Retention

A worked example by Edington (1997) focusing on BAA's approach, has been updated to illustrate the even more acute effects of losing a tenant due to the imposition of tougher regulations on the payment of Business Rates (Commercial Property Taxes) in the UK for empty buildings. The example below is a simple proposition of a multi-tenanted building where the leases come to an end at the same time and a traditional landlord is trying to re-let the offices to the existing tenants on standard leases at the current market rent.

A Worked Example to Demonstrate the Significance of Tenant Retention in Terms of Cash Flow

A traditional landlord has a 10-storey office building with single lettings on each floor of 500 square metres (lettable area) per floor. The property has a current market value of £200 per square metre per annum, the service charge is £25 per square metre per annum and the Commercial Property Taxes (Rates) are £100 per square metre. The leases are coming to an end and the landlord is aiming to re-negotiate with all the tenants.

The gross income per annum produced from the 10 floors on the basis of a new lease at the current market value is therefore:

RENT: 1000 sq m × 10 floors × £200/sq m = £2,000,000

There is a SERVICE CHARGE (Common Cost) which is £25/sq m but it is assumed that expenditure and costs including management costs are balanced.

Suppose that the landlord is not successful in securing lease renewals with three of the tenants as they are not satisfied with the level of service and move to a more contemporary landlord.

If it takes 12 months to find a new tenant the costs to the landlord will be losing rental, being forced to contribute to the void in the service charge and being liable for empty property taxes (at least in the UK):

Loss of Rent	1000 sq m × 4 floors × £200/sq m	= £800,000
Service Charge	1000 sq m × 4 floors × £25/sq m	= £100,000
Empty Property Taxes	1000 sq m × 4 floors × £50/sq m (50%)	= £100,000

In addition, marketing and letting fees are likely to add, say £100,000.
The total cost of this lack of retention for one year will therefore be £1,100,000.

If the landlord had a more contemporary approach and a better relationship with their tenants that led to only *two* tenants not renewing, the gross income will be reduced by only £550,000. Therefore if four tenants are lost the Gross Income will fall from £2,250,000 to £900,000 BUT if only two tenants are lost the Gross Income will fall from £2,250,000 to £1,450,000.

In cash flow terms:
Four defections:

Income (Rent)	1000 sq m × 6 floors × £200/sq m	= £1,200,000
Costs (Service Charge)	1000 sq m × 4 floors × £25/sq m	= (£100,000)
Taxes	1000 sq m × 4 floors × £50/sq m (50%)	= (£100,000)
Marketing/letting fees, say		(£100,000)
Net Income		= £900,000
Two defections:		
Income (Rent)	1000 sq m × 8 floors × £200/sq m	= £1,600,000
Costs (Service Charge)	1000 sq m × 2 floors × £25/sq m	= (£50,000)
Taxes	1000 sq m × 2 floors × £50/sq m (50%)	= (£50,000)
Marketing/letting fees, say		(£50,000)
Net Income (*improved* by £550,000)		= (£1,450,000)
Three defections:		
Income (Rent)	1000 sq m × 9 floors × £200/sq m	= £1,800,000
Costs (Service Charge)	1000 sq m × 1 floors × £25/sq m	= (£25,000)
Taxes	1000 sq m × 1 floors × £50/sq m (50%)	= (£25,000)
Marketing/letting fees, say		(£25,000)
Net Income (*improved* by £825,000)		= (£1,725,000)

In crude cash flow terms, by reducing the defections by two and thereby achieving a 33% increase in renewals (2/6) an increase in cash flow of 61% can be achieved: [(550,000/900,000) × 100]. Reducing the defections to one and thereby achieving a 50% increase in renewals (3/6) an increase in cash flow of 92% can be achieved: [(825,000/900,000) × 100]. Bruntwood, as examined in Case Study 1, achieve three times the average retention rate for the UK.

Reducing Lease Length

According to IPD publications, the average weighted office lease length in 1990 was 23.5 years, and even in 2000 a DETR commissioned study undertaken by Reading

University found that approximately 90% (by value) of new tenancies in 1990 were on lease terms of 20 years or more, the standard review term was for five years and the vast majority of leases were on FRI terms. However, this research also found evidence of a two-tier market with smaller lettings having an average closer to 17 years. By 1992, the IPD rent-weighted average lease length had gradually fallen to just above 20 years but from then onwards it fell more sharply to around 13 years in 1994/1995. Whilst the impact of recession(s) has had an impact at certain times there are underlying reasons of shorter business cycles, more flexible business practices, project and team-based working and different expectations from global companies not used to UK-style institutional leases. Desktop analysis by Hamilton, Lim & McCluskey (2006) revealed that out of 106 leases taken by office tenants in London, Manchester, Belfast and Birmingham from January 2001 to March 2004, the mean lease length was 10.93 years. Research from the annual IPD/Business Property Federation in 2006, recorded the average lease length for premises had fallen to 6.2 years, down from 8.4 in 2001 and that the average small or medium-sized business lease was 5.3 years.

The trend has therefore seen a sharp decline in average lease length although hidden within this data is a multitude of variation with many major traditionally funded projects still demanding longer leases in order to meet investor requirements and at the other end of the scale operators like Bruntwood who offer highly flexible terms and are willing to tear up a lease and start again where appropriate.

Interestingly in the credit crunch challenged market of 2009 some flexibility has been removed from the market, with landlords requiring longer leases due to exceptionally high yields and illiquidity which places demands upon them from investors to manage risk. We managed a project for Dutch Masters students in 2009 which simulated a relocation of a company examining properties in Sheffield and Manchester, in the middle of a severe recession. This company required flexibility as an essential requirement. Most of the agents acting on new developments, with the exception of Bruntwood, were unable to offer short leases. In many cases 20 years were required because the funding arrangements of the developer were approaching their limits in terms of loan to value, a further addition of risk through short leases would have further depressed the capital value resulting in a breach of financial covenant. This may appear counter-intuitive and is a good reminder of how markets and the financial arrangements behind property development have a strong impact upon lease length despite the business requirements of tenants.

Whilst good incentives are given currently for longer leases, we advise you to read Chapter 3 when considering the occupier risks between shorter and longer leases.

Code for Leasing Business Premises

As part of the concerns of the UK government over how the traditional UK institutional lease may make the UK less attractive to contemporary businesses, the government undertook a number of reviews and research into the UK leasing market and encouraged the formation of the Code for Leasing Business Premises in England and Wales 2007 (Joint Working Group, 2007). This is the result of collaboration between commercial property professionals and industry bodies representing both owners (landlords) and occupiers (tenants). The government had threatened on a number of occasions to create legislation which would impose the ideals of the lease code if voluntary uptake was insufficient. Whilst this has not happened, the government keeps a keen watching brief on progress and compliance.

The Code, endorsed by a large number of public and private institutions in the UK, aims to promote fairness in commercial leases. It recognizes a need to increase awareness of property issues, especially among small businesses, ensuring that occupiers of business premises have the information necessary to negotiate the best deal available to them.

Not only applicable to the UK, the Code is endorsed by the European Property Foundation (EPF). As the Code is voluntary not all landlords choose to offer compliant leases and when selecting new premises this may be a criterion to consider. We believe it provides a useful checklist and sets out the minimum businesses should be seeking when selecting a landlord and procuring premises.

The Code consists of three parts:

- a set of 10 requirements for landlords in order for their lease to be compliant with the Code;
- a code for occupiers, explaining the terminology and providing helpful tips;
- a model 'Heads of Terms' the document used to instruct legal advisers in the UK to construct a lease document form, which can be completed and downloaded online.

The Code can be downloaded from www.leasingbusinesspremises.co.uk. The 10 requirements for compliance are set out below and form a useful checklist for negotiations.

1. Lease Negotiations

Landlords must make offers in writing which clearly state: the rent; the length of the term and any break rights; whether or not tenants will have security of tenure; the rent review arrangements; rights to assign, sublet and share the premises; repairing obligations; and the VAT status of the premises.

Landlords must promote flexibility, stating whether alternative lease terms are available and must propose rents for different lease terms if requested by prospective tenants.

2. Rent Deposits and Guarantees

The lease terms should state clearly any rent deposit proposals, including the amount, for how long and the arrangements for paying or accruing interest at a proper rate. Tenants should be protected against the default or insolvency of the landlord.

State clearly the conditions for releasing rent deposits and guarantees.

3. Length of Term, Break Clauses and Renewal Rights

The length of term must be clear.

The only preconditions to tenants exercising any break clauses should be that they are up to date with the main rent, give up occupation and leave behind no continuing subleases. Disputes about the state of the premises, or what has been left behind or removed, should be settled later (like with normal lease expiry).

The fallback position under the Landlord and Tenant Act 1954 is that business tenants have rights to renew their lease. It is accepted that there are a number of circumstances in which that is not appropriate. In such cases landlords should state at the start of negotiations that the protection of the 1954 Act is to be excluded and encourage tenants to seek early advice as to the implications.

4. Rent Review

Rent reviews should be clear and headline rent review clauses should not be used. Landlords should on request offer alternatives to their proposed option for rent review priced on a risk-adjusted basis. For example, alternatives to upward-only rent review might include up/down reviews to market rent with a minimum of the initial rent, or reference to another measure such as annual indexation.

Where landlords are unable to offer alternatives, they should give reasons. Leases should allow both landlords and tenants to start the rent review process.

5. Assignment and Subletting

Leases should:

- allow tenants to assign the whole of the premises with the landlord's consent not to be unreasonably withheld or delayed; and
- not refer to any specific circumstances for refusal, although a lease would still be Code-compliant if it requires that any group company taking an assignment,

when assessed together with any proposed guarantor, must be of at least equivalent financial standing to the assignor (together with any guarantor of the assignor).

Authorized Guarantee Agreements should not be required as a condition of the assignment, unless at the date of the assignment the proposed assignee, when assessed together with any proposed guarantor:

- is of lower financial standing than the assignor (and its guarantor); or
- is resident or registered overseas.

For smaller tenants a rent deposit should be acceptable as an alternative.

If subletting is allowed, the sublease rent should be the market rent at the time of subletting. Subleases to be excluded from the 1954 Act should not have to be on the same terms as the tenant's lease.

6. Service Charges

Landlords must, during negotiations, provide best estimates of service charges, insurance payments and any other outgoings that tenants will incur under their leases.

Landlords must disclose known irregular events that would have a significant impact on the amount of future service charges.

Landlords should be aware of the RICS 2006 Code of Practice on Service Charges in Commercial Property and seek to observe its guidance in drafting new leases and on renewals (even if granted before that Code is effective).

7. Repairs

Tenants' repairing obligations should be appropriate to the length of term and the condition of the premises. Unless expressly stated in the heads of terms, tenants should only be obliged to give the premises back at the end of their lease in the same condition as they were in at its grant.

8. Alterations and Changes of Use

Landlords' control over alterations and changes of use should not be more restrictive than is necessary to protect the value, at the time of the application, of the premises and any adjoining or neighbouring premises of the landlord.

Internal non-structural alterations should be notified to landlords but should not need landlords' consent unless they could affect the services or systems in the building.

Landlords should not require tenants to remove permitted alterations and make good at the end of the lease, unless reasonable to do so. Landlords should notify tenants of their requirements at least six months before the termination date.

9. Insurance

Where landlords are insuring the landlord's property, the insurance policy terms should be fair and reasonable and represent value for money, and be placed with reputable insurers.

Landlords must always disclose any commission they are receiving and must provide full insurance details on request.

Rent suspension should apply if the premises are damaged by an insured risk or uninsured risk, other than where caused by a deliberate act of the tenant. If rent suspension is limited to the period for which loss of rent is insured, leases should allow landlords or tenants to terminate their leases if reinstatement is not completed within that period.

Landlords should provide appropriate terrorism cover if practicable to do so. If the whole of the premises are damaged by an uninsured risk as to prevent occupation, tenants should be allowed to terminate their leases unless landlords agree to rebuild at their own cost.

10. On-going Management

Landlords should handle all defaults promptly and deal with tenants and any guarantors in an open and constructive way. At least six months before the termination date, landlords should provide a schedule of dilapidations to enable tenants to carry out any works and should notify any dilapidations that occur after that date as soon as practicable.

When receiving applications for consents, landlords should where practicable give tenants an estimate of the costs involved. Landlords should normally request any additional information they require from tenants within five working days of receiving the application. Landlords should consider at an early stage what other consents they will require (for example, from superior landlord or mortgagees) and then seek these. Landlords should make decisions on consents for alterations within 15 working days of receiving full information

The Occupier Guide is a very useful document but is not a substitute for legal advice and we would encourage occupiers to use it but also to seek independent professional advice when negotiating a lease.

The RICS Service Charge Code

Dr John Calvert at Loughborough University, sponsored by Property Solutions, has been undertaking longitudinal research into the collection, analysis and reporting of service charges from 2003. The original 2005 research examined 386 buildings and 1788 service charge certificates and the subsequent publication, 'The Trouble with

Service Charges', confirmed that the UK property industry suffered from severe problems with service charges and gave the impetus for the creation of the RICS Code of Practice: Service Charges in Commercial Property.

In the latest (2009) Loughborough report it is noted that 'There are significant performance improvements in many aspects of compliance with the RICS Code of Practice although overall progress against the Code remains disappointing.'

The 2009 report includes a table of key benchmarks within the Code that can easily be measured. It is set out in Table 4.6 and shows that there is considerable work to be done to achieve compliance.

The Code, which is available in editions for England & Wales, Scotland, Northern Ireland and Ireland, should be adopted by all RICS members, and forms an important mechanism for improving the transparency, quality and equity of service charge arrangements. Its principal components are:

- improved communication;
- greater transparency;
- enhanced accuracy;
- no hidden profits;
- value for money;
- duty of care to occupier.

In addition, a separate industry standard for cost headings works alongside the code ensuring consistency by defining cost categories and cost codes under seven cost classes of: Management; Utilities; Soft Services; Hard Services; Income; Insurance; Exceptional Expenditure. Accompanying notes explain each cost code.

TABLE 4.6 Key Benchmarks from 2009 Loughborough Report

RICS Code requirement	April 2007 to 2009 achievement	1998 to 2007 achievement
Budgets must be delivered one month prior to the start of the year	21% (estimated)	5%
Certificates must be delivered within 4 months of the end of the year	46% (estimated)	24%
Management fees must be a fixed cost	33%	18%
Interest must be credited to service charge accounts	28%	16%
Apportionment basis clear	44% (as shown on face of certificate)	77%
Budget accuracy. Budgets should be within 2% of actual costs	15%	14%

The cost codes developed by Jones Lang LaSalle as part of its OSCAR service charge benchmarking system are consistent with the International Total Occupancy Cost Code from IPD, both of which we examine in Chapter 5.

Poorly managed service charges (especially the quality of cleaning and maintenance of common parts) are a frequent cause of disputes between owners and occupiers. The new Code provides the real estate and facilities management (FM) industry with a clear set of recommendations that will increase service transparency between manager and occupier. The report suggests that currently, many tenants of multi-let buildings appear to have little or no idea what they are getting for their money. The quality and scope of service is ill defined and performance measurement is either opaque or non-existent. Again, we recommend occupiers to consult the Code and use it to ensure that any service charges applied to their real estate assets are appropriately managed.

Implications

Thc dysfunctional traditional method of providing property management, separated from services management, will not be adequate to cope with the challenges of quality assurance, value for money, total transparency and simplicity of process demanded by the Service Charge code. Traditional landlords, whose fragmented management structures isolate asset, property and facilities management, create barriers for the introduction of a fully integrated, streamlined property and services management model.

Contemporary companies (as discussed below) have many of the key recommendations of the Code already in place and promote a seamless, transparent and customer focused philosophy.

The Development of Customer Focused Strategies in the UK

In addition to our main case study examining Bruntwood, who have proved the success of customer focus as a strategy through several property cycles, we examine here two pioneers of a customer focused approach and how they have contributed to good practice before examining a contemporary service provider.

BAA: A Focus on Tenant Retention

In the late 1990s Gordon Edington demonstrated how BAA were using customer focus strategically to build BAA as a brand and landlord of choice through customer service. Whilst you might argue that retailers wishing to be a part of the UK airline retail scene had little choice as BAA had a significant control of UK airports, it did contribute significantly to the changing attitude of landlords especially in the retail sector and introduced a number of ground-breaking initiatives including penalty-backed service level agreements.

Edington (1997) defined six key concepts which represent the six building blocks that underpinned the, then, revolutionary property management approach adopted by BAA:

- Defining the customer
- Researching what the customer wants
- Creating a mission for the organization
- Leadership, empowerment, training and communication
- Process improvement and information management
- Measuring success and benchmarking

The first thing to notice is how the list is a departure from traditional definitions of real estate management and contains fundamental modern management concepts including the European Quality model.

Building Block 1: Defining the Customer

This fundamental starting point is actually a stumbling block for property owners: just who is the property customer? Traditionally the investor(s) have dominated as the customer to the property industry and building design, specification and form have been largely dictated by the funds and investors providing finance for a development. Certainly in the UK where the largest proportion of commercial space is built for letting as an investment vehicle the property business has not paid enough attention to the end-user.

Edington believed that customers have a choice and whether that choice is executed on taking a building or in deciding whether to renew a lease, with increased options available to consumers of property landlords should be (and will have to) work harder to retain and attract tenants. In a more sophisticated and mobile market it makes sense for developers and investors to become more customer focused to reduce the risk for their investor clients.

Block 2: Researching What the Customer Wants

Historically, tenants are rarely treated as proper customers. Having signed a 25-year lease in the UK, landlords want little more to do with a tenant other than to ensure that their income stream is protected. The lack of post-occupancy evaluation, a common FM tool, means that there is an absence of a vital feedback loop for developers to learn from the reaction of their tenants (customers) to the buildings they are occupying. Listening to this feedback could result in a continuously improved evolution of design, specification and features to improve the lettability of future products. Listening, from a management perspective, as to the running of the building, day-to-day operations and feedback from customers, supplies and visitors is another vital, but often sadly lacking activity.

BAA, Bruntwood (see Case Study 1) and other enlightened companies regularly interface with their customers and collect both individual tenant and estate based data and responses with a genuine commitment to act upon the feedback and strive for continuously improving the quality of the services they provide.

Building Block 3: Creating a Mission for the Organization

Edington (1997) described this as being one of the most critical aspects of the process. Unless the desire to continuously improve performance and offer a superior and excellent service is not embraced by everyone as the company culture then the rhetoric will not be the reality. In many organizations mission statements are merely rhetoric and if not believed by all of the staff who interface with the customers they are often ineffectual.

Building Block 4: Leadership, Empowerment, Training and Communication

Leadership is another undervalued and sadly lacking commodity in many businesses. Studies of quality systems and models, and those companies that strive for international excellence have demonstrated that leadership is one of the key success factors.

Bureaucracy is a barrier to customer response and a relaxation of traditional controls and the empowerment of staff to act (within boundaries) to situations without the need to obtain permission is an important part of the service culture. At Bruntwood, property service managers are empowered to act on behalf of the company; they do not have to confer with landlords or funds and can simply get the job done.

Building Block 5: Process improvement and Information Management

This block applies business process re-engineering to re-focus the operations of the real estate organization and to seek efficiency and effectiveness through radical re-shaping of operations.

The re-engineering process can also be a catalyst for examining and collapsing traditional boundaries in organizations and eliminating the 'not my department's responsibility' attitude which is often a serious distractor from customer service.

Building Block 6: Measuring Success and Benchmarking

In order to state that improvements have been made a benchmark is needed against which the improvement can be measured.

BAA has both internal benchmarks to compare property and customer servicing amongst its airports and terminals within large airports such as Heathrow. However, external benchmarks comparing with the success of other world class businesses are also encouraged to promote long term improvements in standards to match the world class expectations of its customers. This provides a management 'stretch' or long term improvement goal. An example of this is given by Edington (1997).

Edington discusses a visit to the United States to benchmark best practice, where it was discovered that managers of major high rise buildings were committed to resolving the majority of property faults and minor maintenance – for example, replacing light bulbs within 15 minutes. This continues to be a core deliverable for forward thinking organizations such as the Oxford Properties Group in the USA.

BAA at the time had a target of resolving only 80% of faults within 48 hours. This provided the management stretch and the company set itself and achieved a target of resolving 95% of property faults within 4 hours. In addition, BAA introduced a revolutionary penalty backed guarantee concept which, at the time, and sadly even today, made conventional landlords gasp. This scheme ensured compliance with the targets, such as that described above, by offering penalties, including foregoing rent, if service delivery guarantees were not met. In many other sectors this is quite commonplace; where service level agreements are not met penalties, previously agreed in the SLA are applied. However, they remain uncommon in the property sector.

It is interesting to note how these building blocks resonate with the REAL SERVICE benchmarking building blocks discussed in Case Study 1: Bruntwood.

Arlington Properties: Innovation in Service Delivery

Howard Bibby was the architect of many innovative and creative approaches to service delivery for the Arlington Group. Arlington is now part of the Australian international property group Goodman following its acquisition of Arlington and a number of other European property development and management operations between 2005 and 2007. Whilst we are sure that many of the innovations developed by Arlington remain within the Goodman operations, the scale of the operation makes it more difficult to report, so we have concentrated on the Arlington based initiatives.

In 2003 Arlington Securities Plc won the RICS Property Management Awards for its customer focused approach strategy, largely attributable to Howard Bibby who was the Managing Director of Arlington Business Services from 1997 to 2005. The 'Arlington Way' was based on a *'challenge to continually add value to our customer's business'*. This was achieved by:

- providing and managing effective working environments that enable companies to enhance their business performance;

- designing and delivering integrated service solutions that are 'quicker, simpler, better';
- ensuring value for money from the right balance of quality, risk and cost;
- delivering consistent service quality in terms of performance and customer satisfaction.

This in turn was underpinned by a set of values which ensured that Arlington would innovate and differentiate from other service providers and provide genuine customer service. The values included the following statements:

- We keep everything simple and hassle free.
- The customer can trust us to get tasks right, first time, on time and in full.
- We want long term sustainable relationships – we deal ethically with our customer.
- As individuals we will personally commit to own and solve customer problems.

The mechanics behind the strategy relied on principles of transparency and continuous feedback. Arlington was unique in publishing customer satisfaction surveys and creating online satisfaction 'performance dashboards' that tenants using its management portal could participate in and observe. Actual Service Level Performance was measured every month automatically through their online Service Desk. All complaints, planned and reactive tasks were pre-assigned a response and completion time. Tenant retention (like Bruntwood) was way above national averages with retention rates of up to 90% on some developments for standalone lease contracts.

In addition, Arlington pioneered the introduction of non real estate service staff, including those from the Hotel and Leisure Industry who fully understood customer service and could work alongside property professionals dealing with the transactional side of the business. Arlington again backed up its service promise by linking performance related pay to salary and bonus levels for customer-facing staff.

One measure of the success of this approach is that in 2004 Arlington achieved a 90% average customer satisfaction rating (compare that to the 2009 UK Occupier satisfaction index results) and an average monthly service desk performance of 90%.

Fasset Management: Flexible Attitudes Towards Leases

Fasset Management has core values based on being very customer focused and contributing to the economic development agenda of the regions it operates in. It has been successful in working with corporate occupiers to turn 'legacy' sites no longer required by an organization from a cash drain to viable business solutions offering either a return or capital receipt. Flexible attitudes towards leases are viewed as the way to ensure a high level of occupancy and tenant retention. The increasing cost to landlords of vacant premises (particularly through the recent amendments to empty business rates legislation) is as set out in the worked example in this chapter. Combined with the impact of the recession, this has meant that preserving the cash flow through the retention of existing tenants is critical. Fasset management has adopted a contemporary approach which is primarily driven by obtaining high occupancy levels in its managed premises through a flexible attitude. Fasset has managed to achieve occupancy rates in excess of 90% throughout its portfolio (www.fasset.co.uk).

Fasset's first and most recognized project is Langstone Technology Park. Originally a disused and redundant office premises previously occupied by Xyratex (a subsidiary to IBM), Fasset's management team looked to innovative ways of securing tenants and thus increasing the landlord's investment value. Traditionalists valued the building at site value due to the awkwardness of the design and large amount of apparently unusable space and saw little scope for finding a suitable tenant. Fasset, however, decided to base its management approach on an American Science Park model and focused on providing a similar fully serviced environment, tailoring space and occupancy packages with a customer focused approach. Fasset's success has resulted in the eventual sale of the site for investment purposes with bidders battling it out to a final sale of £54 million reflecting an initial yield of 7% and a 150% valuation increase over the 5-year management period.

Fasset provides flexible lease terms and quality, fully serviced facilities for companies looking for a low risk, all inclusive occupancy package that allows for minimum initial investment and maximum flexibility. With offices designed to meet the needs of a broad market Fasset can provide tailored packages of real estate and FM services to best meet the needs of the client.

Fasset has been able to appeal to a far greater sector of the market, where tenant turnover is no longer of greatest importance. With services provided at typically 15% below the average cost of tenants sourcing them for themselves, it is an attractive proposition that drives tenant retention.

Conclusion

Real estate decisions are increasingly recognized as key features in the success or downfall of businesses. This has of course been intensified during the current economic climate. Occupiers are quickly realizing the implication of inflexible property portfolios and their inability to meet rapidly changing business needs. When coupled with an inflexible traditional landlord who may frustrate the implementation of changes designed to promote productivity or increase sustainability, we believe the choice of a more customer focused landlord is clear.

Summary Checklist

1. Undertake a review of your lease to examine whether it complies with the Code for Leasing Business Premises (Joint Working Group, 2007). If you are entering a lease or renewing your lease, benchmark it against the Code.
2. Review your service charge provisions against the RICS Code of Practice (2007).
3. Review your level of customer service against those presented in this chapter and/or Bruntwood in Case Study 1. Would your business benefit from the flexibility, customer service and other benefits offered by landlords that offer a more contemporary approach?
4. Review the utilization of your business space. If it is volatile, would a landlord with a more flexible approach help to manage the volatility?
5. Does your landlord measure its customer service with a view to improvement? Is the lack of service affecting your business, in which case on renewal of your lease you should consider carefully what alterative options are available.

References

Bibby, H. (1999). The Arlington Way. *Premises & Facilities Management*, January.

Calvert, J., Barrass, C., Morgan, A., & Reimer, J. (2005). *The 2005 update: The trouble with service charges.* www.property-solutions.co.uk. Accessed 12.11.09.

Calvert, J., Barrass, C., Morgan, A., & Reimer, J. (2009). *The Loughborough Report 2009.* www.property-solutions.co.uk. Accessed 12.11.09.

Crosby, N., Gibson, V., & Murdoch, S. (2003). UK commercial property lease structures: Landlord and tenant mismatch. *Urban Studies, 40*(8), 1487–1516.

Crosby, N., Gibson, V., & Oughton, M. (2001). *Lease structures, terms and lengths: Does the UK lease market meet current business requirements?* Reading: University of Reading/RICS.

DETR (Department of Environment, Transport and the Regions). (2000). *Monitoring the Code of Practice for commercial leases.* Research report by the University of Reading. London: DETR.

Edington, G. (1997). *Property management: A customer focused approach.* Houndmills: Macmillan.

Hamilton, M., Lim, L., & McCluskey, W. (2006). The changing pattern of commercial lease terms. Evidence from Birmingham, London, Manchester and Belfast. *Journal of Property Management, 24*(1).

Joint Working Group on Commercial Leases. (2007). *Code for leasing business premises in England and Wales 2007.* http://www.leasingbusinesspremises.co.uk. Accessed 12.10.09.

RICS. (2007). *RICS Code of Practice: Service charges in commercial property.* www.servicechargecode.co.uk. Accessed 12.10.09.

Scarrett, D. (1983). *Property management.* London: E & FN Spon.

Chapter 5

Performance Measurement and Benchmarking

Introduction

CREAM requires a thorough understanding of the performance of an organization's assets – measured both internally and externally, in comparison with other organizations in the same sector.

In this chapter we aim to explore the practical side of benchmarking and real estate performance measurement. We look at how both can be used to identify problems in a corporate or occupier portfolio, compare performance both internally and externally, and support strategic business decision making. We also evaluate a number of published benchmarks, which examine the use of space, occupancy costs and service charge costs. In this chapter we focus mainly on quantitative benchmarks; we examine qualitative measures in Case Study 1: Bruntwood.

Some Definitions of Benchmarking

Benchmarking is a broad and complex area of practice, so we aim to keep it simple, focusing on the key indicators and published benchmarks that can support effective corporate real estate management. Its origins are not agreed but many point to land surveying and the use of poles and angles related to a surveyor's desk or 'bench'. It is now used as a widespread term for both a process and as a relative measure.

It is important to contrast performance measurement and benchmarking. Performance measurement is usually the observation or measurement of performance within an organization. For example, a company may calculate the total occupancy cost of space per person used within its offices, and monitor it over time. However, this only indicates the company's performance in isolation. Published information such as the DTZ Global Office Occupancy Costs Survey could be used as a crude benchmark to compare the organization's occupancy costs against an average for a particular location. So, for example, if the organization is based in Boston, USA

Corporate Real Estate Asset Management. ISBN: 978-0-7282-0573-4

and established that its total occupancy costs were measured as $13,750/workstation it would know that it compares favourably against the DTZ published figure of $15,640 (2009). However, this type of crude benchmarking would mask significant problems of comparability, measurement and standardization which are overcome by some of the more sophisticated approaches we describe in this chapter.

A good example of a benchmarking in the real estate sector is the Investors Property Databank (IPD). IPD provide both generic and bespoke benchmarking data for property investors based upon a very comprehensive dataset provided by a high percentage of landlords in the UK, and increasingly in other international locations.

IPD produces, for example, a number of performance measures for the UK in terms of total, income and capital returns by property sector. If we take the IPD Index results to December 2008 it indicates that the total return for office property is −22.4%. This figure is based upon data from 11,214 properties, valued at £130 billion. Here we have a benchmark, used by funds to compare their performance against what is a very large percentage of funds under management. As the numbers quoted above are from a recessionary period, they are negative. However, in periods of growth, say the return was 6.3%, fund managers would be expected to outperform the relative IPD index benchmark by a set percentage. We also see here a benchmarking process in action. IPD uses a specific model with well published and transparent methodology to ensure users engage with their benchmarks in a consistent way. Funds themselves are using a benchmarking process to compare their performance with other funds; they can choose to set targets based on the average performance of a sub-set of similar investments or they could choose to outperform the best performing funds within the IPD dataset.

Definition of Benchmarking as a Process

> *benchmarking is the process of identifying, sharing and using knowledge and best practices*
>
> Maire, 2002
>
> *It represents a management tool for enhancing an organization's performance and competitiveness.*
>
> Johnson & Scholes, 1999

These definitions reflect the origins of management benchmarking, frequently attributed to Rank Xerox who used the technique to examine Japanese competitors' processes against their own in 1979. As Zairi (1998) points out:

> *Rank Xerox, faced with Japanese competition which began selling copiers for less than they could manufacture them for, had to find out how they did it to survive.*

In business, benchmarking is often about comparing processes to enhance performance. The classic tale of how SAS, the Scandinavian airline, used an analysis of F1 racing pit stops to enhance their aircraft turnaround times is a good example of how examining processes of a similar nature but from a different sector can also be used to gain strategic advantage.

However, in a real estate context, benchmarking tends to be focused on analysis of performance and seeks to examine tangible measurements such as sales per square metre or total occupancy costs per square metre rather than more abstract business processes.

Benchmarking is frequently divided into six or seven types in the main literature. Here we have briefly reviewed each type and provided examples:

- ***Strategic Benchmarking*** – aimed at improving a company's overall performance by studying the long term strategies and approaches that helped the 'best practice' companies to succeed. It involves examining the core competencies, product/service development and innovation strategies of such companies – the classic benchmarking approach adopted by SAS.
  - **Property example:** Bruntwood, discussed in more detail in Case Study 1, have utilized comparisons with the hotel industry and Lexus cars as a way of improving customer servicing for their tenants.
- ***Competitive Benchmarking or Performance Benchmarking*** – used by companies to compare their positions with respect to the performance characteristics of their key products and services. Competitive benchmarking usually involves companies from the same sector.
  - **Property example:** Retailers will use this approach to compare their performance, for example sales per square metre, against their competitors.
- ***Process Benchmarking*** – used by companies to improve specific key processes and operations with the help of best practice organizations involved in performing similar work or offering similar services.
  - **Property example:** BAA examined other industries' and USA based operations' service level agreements to improve service delivery and turnaround times of lease management operations.
- ***Functional Benchmarking or Generic Benchmarking*** – used by companies to improve their processes or activities by benchmarking with other companies from different business sectors or areas of activity but involved in similar functions or work processes.
  - **Property example:** A number of customer focused landlords have examined the hotel industry to introduce more customer focused lease management and the provision of additional services to tenants, including concierge services in office buildings.

- ***Internal Benchmarking*** – involves benchmarking against an organization's business units of the company situated at different locations. This allows easy access to information, even sensitive data, and also takes less time and resources than other types of benchmarking.
  - **Property example:** Retailers such as Boots the Chemist will use internal rents to calculate occupancy costs and compare this across the portfolio; they may also examine the sales to costs on a square metre basis across their UK high street stores.
- ***External Benchmarking*** – used by companies to seek the help of organizations that succeeded on account of their practices. This kind of benchmarking provides an opportunity to learn from high-end performers.
  - **Property examples:** Companies may want to examine their performance against others; there are several examples of this which we discuss in this chapter. IPD measures investment performance of the majority of portfolios in the UK and increasingly internationally. Portfolio managers can compare their own performance against that of the IPD universe or individual subsets. This can also be done for occupancy costs, using IPD's Space Code and Occupancy Costs benchmarks or that produced by Actium Consult.
- ***International Benchmarking*** – involves benchmarking against companies outside the country, where there are very few suitable benchmarking comparisons within the country; or where international comparisons, for example sales, are required.
  - **Property example:** IPD can be used to examine international portfolio composition and performance and DTZ produce an annual table of occupancy costs in 137 business districts in 49 countries. Cushman & Wakefield compare European City performance to help corporate location decisions in their annual European Cities Monitor (see Chapter 7).

Benchmarking in Practice

What Is Measured?

It is useful to start with an examination of some concrete examples of what is measured by organizations to help them monitor and improve performance and what data are available for organizations to make comparisons against, usually referred to as benchmarks. Later we will investigate some detailed examples to show how performance and benchmarking can be used to identify positive and negative aspects of a portfolio; as a tool to improve both real estate and business performance and to help formulate strategic decision making. Set out in Table 5.1 are examples of performance measures and benchmarks used in practice.

TABLE 5.1 Examples of Real Estate Performance Measurement Categorized into Four Significant Areas: Cost, Efficiency, Utilization and Quality

Measurement group	Example of performance measured	Benchmark utilized in practice to compare performance against
Cost	Occupancy costs per workstation	United States General Services Administration Cost per Person model DTZ Total Office Occupancy Costs IPD Occupancy Costs Databank
	Total real estate occupancy costs per square metre	IPD Space Code Actium Consult Total Occupancy Cost Survey
	Reduction from headline rent to net effective rent (for 50 locations in the UK)	Actium Consult Total Occupancy Cost Survey
	Service charge costs per square metre and by component, e.g. security	Jones Lang LaSalle OSCAR
Efficiency	Amount of space per person	IPD Space Code DTZ Total Office Occupancy Costs Actium Consult Total Occupancy Cost Survey
	Amount of vacant space within a portfolio	Measured by specific organizations such as BT Measured by IPD Occupiers
Utilization	Percentage of office space occupied on a daily/weekly basis	Measured by specific organizations no generic published benchmarks
	Utilization of meeting rooms on a daily/weekly basis	Measured by specific organizations no generic published benchmarks
Quality	Tenant satisfaction indicators	Measured by customer focused landlords such as Bruntwood. External benchmarking provided by Real
	Calls made to landlords to report problems with a building (a proxy indicator for building quality and performance)	Measured by companies such as LloydsTSB; no generic published benchmarks

Performance Measurement in Practice

It is worth noting that many businesses will seek to tailor their performance measurement to their business operations. This helps them understand real estate commitments in the context of their business but may make it difficult to obtain published comparative benchmarking data.

Set out below are some real examples of both generic and bespoke performance measurement used by leading organizations, from a study by Bon, Gibson & Luck (2000).

- cost to income ratio by property (NatWest Bank)
- property cost per passenger kilometre flown (British Airways)
- property cost per tonne-kilometre flown (British Airways)
- occupancy cost per square metre (British Gas)
- occupancy cost per employee (HSBC)
- number of help desk calls per square foot (LloydsTSB)

The 'Apples and Pears' Problem

All benchmarking exercises are often plagued by inconsistencies with data, making comparability difficult and in some cases meaningless. This is frequently referred to as the 'apples and pears, not apples and apples' problem. In real estate this is also a major issue as property benchmarking may be comparing properties which vary in terms of:

- **Measurement** – buildings are measured in different ways across the world making for example the calculation of building efficiency (the ratio of gross built space to net usable space).
- **Occupation type** – freehold, leasehold, length of lease etc.
- **Depreciation/de-capitalization** – different rates may be used to allow for depreciation or de-capitalization of equipment such as IT or furniture over a time period when calculating occupancy costs.

In order to explore this issue we intend to examine these three areas in greater detail:

Measurement

Many real estate benchmarks are related to area – for example total occupancy costs per square metre, and building efficiency. The problem is that the way buildings are measured varies dramatically across Europe and even more so across the globe. It is essential therefore if an organization is undertaking benchmarking that a consistent approach is adopted. There are many definitions of measurement and codes of measuring practice. We start by examining the Royal Institution of Chartered Surveyors (RICS) code of Measuring Practice. The 6th edition published in 2007 clearly states that it is intended for use only in the UK and therefore whilst useful if benchmarking in the UK only it does not address pan-European or global measurement practices. The RICS code defines three core definitions:

GEA (Gross External Area) – the area of a building measured externally at each floor level.
GIA (Gross Internal Area) – the area of a building measured to the internal face of the perimeter walls at each floor level.
NIA (Net Internal Area) – the usable area within a building measured to the internal face of the perimeter walls at each floor level.

Each of the definitions has sets of specific inclusions and exclusions to cover balconies, mezzanine floors, lifts and outbuildings.

Accurate and consistent measurement is critical but the biggest problem arises with the definition of the word *usable* when calculating net internal areas. This is a common calculation and it underpins many benchmarking and performance calculations. The efficiency of a building may be benchmarked from a landlord's perspective, i.e. how much floor space is usable and therefore rentable compared to the built area: the higher the ratio the more efficient the space. It may also be critical in the calculation of service charges, space utilization ratios and productivity – sales per square metre of usable retail space or costs of staff per square metre of net usable area.

The problems occur when companies try to benchmark across national boundaries: for example in Europe there are many standards that do not have a consistent basis of measurement. In addition to the RICS Code for the UK there is NEN 2580 used in Belgium and The Netherlands and DIN 277 used in Germany, Austria and Switzerland. In the United States the Building Owners and Managers Association (BOMA) Floor Rentable Area (FRA) standard is used. Research by Jones Lang LaSalle (2005) demonstrated the significant differences between countries and examined the variation in leased area from the BOMA FRA to locations in Europe.

Their findings suggest that deviations from the BOMA FRA would be:

- UK −13.6%
- France −2.5%
- Norway, Turkey and Israel +7.2%

When comparing against the RICS NIA they suggest the situation is even more diverse. Deviations from the RICS NIA would be

- USA 15.7%
- Belgium 21.2%
- Norway 24.1%
- Sweden 7.3%
- Germany 12.2%

Using local measurements to calculate real estate benchmarks across a portfolio could therefore create a significant 'apples and pears' issue. The study indicates, for example, that a total difference in area of 24.1% may occur when comparing 'usable' areas between the UK and Norway without adjustment, this would be highly misleading.

To combat this problem a new code for measurement in Europe is being proposed as part of a major initiative to allow pan-European FM benchmarking, known as EN 15221 and colloquially called the *'FM Six Pack'*. This European Commission initiative includes a number of measures designed to improve efficiency, effectiveness and competitiveness of FM. The new standard is designed and developed from the user/renter perspective not from the standard investor/developer one. The new system will change the measurement in many countries; in The Netherlands it will mean more square metres are included in the rentable area but ultimately, if widely adopted it will mean 'equal square metres' across Europe and has the flexibility in time to be made consistent with the BOMA FRA measurement practice.

Occupation

When comparing properties in a portfolio for benchmarking purposes it is possible that some properties may be freehold and others leasehold. Therefore, in calculating, for example, total occupancy costs across a mixed retail portfolio, there is a problem as rent will be a major component of the leasehold property but not within the freehold properties. It is possible, but not desirable, to use an alternative measure to reflect the real estate costs such as mortgage finance if the property has a loan, but this creates a classic 'apples and pears' issue.

Many retailers, including Boots the Chemist, have for a number of years used a comprehensive and robust system of *'internal rents'* on their freehold properties. Creating an internal rent, equivalent to market rent, can be used for performance measurement and benchmarking purposes and removes the 'apples and pears' issue. Companies who do not follow this strategy may find that their freehold properties demonstrate higher performance which is not fully justified because a market rent has not been included within the benchmarking calculations. Boots, in calculating, for example, a store-wide benchmark of sales compared to total occupancy costs, by using internal rents for all its properties whether leasehold or not, can perform a consistent and meaningful analysis.

Occupation will also have a significant impact upon the way in which the assets and liabilities of ownership are accounted for in the company accounts. This is explored in more detail in Chapter 3.

Capitalization

In calculating total occupancy costs of offices it may be necessary to examine not only revenue (periodic) based items such as rent, service charge and property taxation but also capital (one-off) items such as furniture, IT infrastructure, telephone systems etc. It could be argued that the de-capitalization should be related to the period of occupancy (lease length) as it is likely that the occupier will use this as an appropriate measure. However, this will vary across a portfolio leading to an 'apples and pears' issue. To combat this some of the leading benchmarking systems have adopted standardized recommended de-capitalization periods related to a combination of industry practice and/or accounting principles. These tend to follow a straight line depreciation approach and may be shorter than standard accounting principles to allow for technical obsolescence etc. For example: IPD Occupiers (formerly OPD) uses the following standardized approach:

Telephone and reprographics 3 years
Acquisition 5 years
Furniture and equipment 5 years
Adaptation and fit out 7 years
Major repairs 10 years

IPD Occupiers is a very complex and comprehensive occupancy cost benchmarking system. It allows members of the benchmarking 'club' to share data and thereby creates a benchmarking system that can be tailored to compare an organization's own occupancy costs with others across a sector or location.

Criticisms of Benchmarking

Although benchmarking is very effective overall, it does have limitations. The main problem with benchmarking is that it may focus too much on data not the processes behind that data. Other criticisms include:

- benchmarking practices are retrospective, they measure company history rather than incorporating current trends and projections, therefore they are of limited assistance to an organization aiming to achieve a competitive advantage;
- no one outside the 'best practice' companies really knows what the 'best practices' actually entail; what is seen from the outside is the results of best practice NOT best practices themselves;

- copying competitors' approaches produces inherently conservative improvement and may inhibit innovation; copying IPD portfolios can produce a 'herd' investment mentality;
- benchmarking cannot be applied to measure qualitative performance indices such as brand strength;
- organizations need to determine what will work for them; no two organizations have the same resources or capabilities therefore working towards their own goals is what really matters and benchmarking can support this;
- in the past, over-reliance on readily available, aggregated data rather than customized benchmarks has produced inconsistencies because assessing performance relative to industry averages is like comparing apples to a basket of mixed fruit.

A Slow Take-up in the Real Estate Industry

The literature is littered with evidence that the real estate industry has been slow to take up performance measurement and benchmarking. Avis & Gibson, of Reading University, have carried out a number of major studies in the UK and throughout Europe in collaboration with some major UK occupiers, including British Telecom and the management consultants Ernst & Young.

The 1995 study produces some very disturbing facts:

- only 13% of the 155 organizations surveyed had property representation at Board level, despite property being one of the three main costs for over 50% of organizations;
- 40% of the organizations were not satisfied with their information systems.

In 1995, in answer to the question 'does your organization have specific performance indicators for property?':

- 37% of all organizations said yes;
- 59% said no.

less than a third specifically tracked return on assets, costs per area and property costs as a proportion of turnover; very few monitored property costs per employee or employee's satisfaction with the space provided.

In the 1999 study, 60% of respondents felt that their information systems were not able to provide reliable decision making data. In the same study a very worrying picture emerged of the lack of integration of corporate data and objectives with the property function – again 60% of the organizations in the survey indicated that availability of data on turnover/income, corporate cost base, corporate assets and human resources was inadequate.

On Arthur Andersen's website (2002), you could find the following: 'Corporate real estate should be viewed as a strategic asset that can impact corporate performance.' Around this time the management consultants and accountants seemed to be leading the way in suggesting that they were best placed in their role as understanding the business to integrate real estate in performance measurement and benchmarking.

Is the Situation Different Now?

In the last ten years there have been a number of public and private sector initiatives which have driven forward benchmarking. In the public sector the picture is quite confusing. In 2000 the Audit Commission published *Hot property: Getting the best from local authority assets*, and at that time Best Value reporting of local authorities required mandatory monitoring and reporting of a range of real estate related measures. However, more recently in the UK a single set of 198 National Indicators for Local Authorities and Local Authority partnerships have replaced Best Value performance indicators and these only include sustainability measures related to real estate.

In 1997 the Chartered Institute of Public Finance & Accountancy introduced the National Best Value Benchmarking Scheme. Using a series of Web-delivered modules members can enter information to templates and receive instant comparative data and benchmarking information. Adopted by many authorities in Scotland and 13 London boroughs, it provides a useful route to generating not only information previously required by Best Value measures, but a range of performance measures and benchmarks critical for effective management of real estate.

Under the Estates & Property Module it provides relative performance indicators including:

- staff costs as a % of rental income
- rent arrears as a % of rental income
- % of rent reviews served on time
- annual growth of portfolio rent roll.

Probably the most significant is the Office of Government Commerce (OGC) High Performing Property Initiative (2007) which is being delivered in partnership with IPD. This is considered in more detail later in this chapter as the approach has valuable lessons for both public and private sector organizations.

In the private sector there appears to be much greater awareness of the opportunities for real estate to be used as a strategic asset and evidence that more companies, especially large companies, are using benchmarking data. The growth of occupier benchmarking data, including Actium Consults Total Occupancy Cost Service (TOCS) and the launch of the IPD Space Code, demonstrates the increasing availability of data and interest. However, in 2002 the RICS report *Property in business: A waste of space* still claimed:

> *It is surprising that many businesses do not have an accurate assessment of the property costs they face …*

In our experience there appears to be a broad spectrum of utilization of performance measurement and benchmarking. Some companies have embraced it and use it strategically, in others there is still much work to be done and many companies do not have a comprehensive and consistent approach to monitoring their assets in a systematic way. We concur with authors such as Nourse (1994), who identified in his research a link between the number of performance measures and benchmarks employed and the tighter links between business strategy and real estate operations.

Even where companies subscribe to proprietary benchmarks it does not necessarily mean that organizations are using it and if they are that it is being used in an effective and systematic way. At the end of this chapter we review how benchmarking can be applied strategically to avoid dangers of simple, one-dimensional comparisons with published benchmarks.

Benchmarking in Action 1

Space Standards – Preoccupation with a Single Standard

There have been a large number of attempts to define a space per person benchmark and most recently the IPD has produced a useful table in their *IPD Occupiers efficiency standards for office space: A report to Office of Government Commerce*, November 2007, reproduced below as Table 5.2.

In this document they advocate a space benchmark of 12 square metres per person for government offices. We have concerns around using a single measure and argue, in Chapter 6, that space should be seen as an essential component of productivity. The diversity of space, from hot desks to directors' clubs, suggests it is meaningless to work with a single figure other than for the crudest of assignments. Nevertheless, there is a preoccupation with calculating this crude measurement and perhaps not surprisingly due to the high cost of space and the introduction of more innovative working arrangements the space per person has declined over the years.

The problem therefore with these headline numbers is that they hide the diversity of organizations and the working practices within them. Whilst the latest IPD report for OGC is useful in providing a driving force to reduce occupation densities in the public sector from an average of 15 sq m per person towards the 'magic' number of 12 sq m per person they may encourage insufficient focus on innovation in the workplace, such as working beyond walls, as discussed in Chapter 6.

TABLE 5.2 A Comparison of Space Standards

Key occupancy space per person benchmarking reports source	Space per person benchmark	Comments	Publication date
Roger Tym & Partners[1]	17.9 sq m	Based on study of the South East	1997
Gerald Eve[2]	16.3 sq m (range: 10.6 sq m to 19.7 sq m)	National survey, cross-sector, with large sample	2001
Arup Economics & Planning[3]	City, 20 sq m Business parks, 16 sq m General offices, 19 sq m	Arup presented their figures in gross rather than net lettable	2001
TOCS[4]	14 sq m	Up to 12.5 sq m in the IT sector	2003
DTZ[5]	20 sq m	Study of the SE, excluding London	2004
BCO[6]	14 sq m (range: 12 sq m to 17 sq m)	National guidance based on understanding of best practice	2005
IPD[7]	14.8 sq m	0.3 sq m, 130 building sample of the government estate	2005
Roger Tym & Partners, et al.[8]	16.2 sq m (range: 14.4 sq m to 20.6 sq m)	London study, large sample	2006
Greater London Authority[9]	16.3 sq m	Rising to 13.9 sq m in forecasts of future standards	2007
IPD[10]	14.5 sq m	Based on a sample of 375 offices, 95,000 people and 1.4m sq m	2007

[1]Roger Tym & Partners (1997). *The use of business space: employment densities and working practices in South East England.* London: Serplan.
[2]Gerald Eve (2001). *Overcrowded, underutilized or just right?* London: Gerald Eve.
[3]Arup Economics and Planning (2001). *Employment densities: A full guide.* London: AEP.
[4]Actium Consult & CASS Business School (2003). *Total office cost survey.* London: Actium.
[5]DTZ (2004). *Use of business space and changing working practices in the South East.* London: SEERA.
[6]BCO (2005). *BCO guide 2005: Best practice in the specification for offices.* London: BCO.
[7]IPD Occupiers (2005). *Property benchmarking project OGC/51: Final report for pilot phase.* London: IPD.
[8]Roger Tym & Partners, Ramidus Consulting & King Sturge (2006). *The use of business space in London.* London: RTP.
[9]Harris, R., Chippendale, D., Cundell, I. & Jones, S. (2007). *London office policy review 2007.* London: Greater London Authority.
[10]IPD (2007). *OGC Property benchmarking 2006 report.* London: IPD.

The new IPD Space Code (2008) has been developed as *'a framework for the collection, measurement and analysis of floor space information'*. It is designed as a global standard of measurement which enables companies to benchmark their property portfolios universally either externally or internally for global companies. The aim of the Space Code is to aid companies in rationalizing the use of space within their properties. IPD has surveyed large occupiers and found a broad consensus that more effective use of space is the primary means of contributing to better organizational efficiency.

IPD identifies certain benchmarks which are most important for increasing space efficiency and reducing total occupancy costs:

- reduce space per workstation;
- increase people per workstation;
- increase usable : total floor space ratio.

Whilst these benchmarks place heavy emphasis on reducing the space necessary to undertake operations, IPD recognizes the importance of contextual environmental factors in determining the overall productivity of a workspace. The report states the need to allocate larger spaces for employee interaction which is deemed to improve individual productivity. Furthermore, IPD emphasizes a need for attractive and amply sized client-facing areas. The retention of clients will have significant implications on a company's financial outlook which could outweigh the benchmark of total occupancy costs and therefore the two need to be taken in conjunction.

De Jonge et al. (2008) suggest that benchmarking measures such as cost per square metre focus only on the efficiency of real estate rather than the effectiveness it has on other factors:

> *The impact of real estate cannot be isolated from the impact of other variables such as capital, technology, human resources or ICT, and from the external context.*

For example, desk sharing will reduce the amount of space allocated for workstations and therefore directly reduces total occupancy costs. However, a large interaction area for staff in a well-designed building may increase employee satisfaction and could therefore indirectly reduce costs through reduced staff turnover. Benchmarking targets need to be met in line with other environmental variables specific to a company in order to fully address their real estate needs. In some situations it may be more financially beneficial to maintain high employee or customer satisfaction at the expense of higher total occupancy costs.

We have some concerns about how some of these approaches may not sufficiently allow for the wide variations in work styles and workplace settings and the sophistication of knowledge based companies. We agree with Osgood (2002a), who questions the 'viability of simple comparisons like sq.ft per person, unless you understand the core business and real estate components that comprise the statistic'. What is needed is an approach that blends business and real estate performance and strategy and reflects the complexities and subtleties of human workspaces. We explore this in more detail at the end of this chapter.

Benchmarking in Action 2

The UK Office of Government Commerce, Working in Partnership with IPD

The United Kingdom's central government estate comprises over 300 individual property centres, 13 million square metres of space and annual costs of £6 billion. Each department within this vast estate is responsible for procuring and managing its own property portfolio. Many of these departments have their own systems in place to monitor the overall efficiency and effectiveness of their estate, however there was no overall plan to monitor this across the entire central government estate.

'The Office of Government Commerce (OGC) is an independent office of HM Treasury, established to help Government deliver best value from its spending' (www.ogc.gov.uk). It currently aims to reduce government spending on its civil estate by an ambitious amount of between £1 billion and £1.5 billion by 2013.

The OGC has six main goals:

- delivery of value for money from third party spend;
- delivery of projects to time, quality and cost, realizing benefits;
- getting the best from the government's £30 bn estate;
- improving the sustainability of the government estate and operations, including reducing carbon emissions by 12.5% by 2010–11, through stronger performance management and guidance;
- helping achieve delivery of further government policy goals, including innovation, equality, and support for small and medium enterprises (SMEs);
- driving forward the improvement of central government capability in procurement, project and programme management, and estates management through the development of people skills, processes and tools.

The absence of an all-encompassing estate management strategy, consistent monitoring and performance measurement led to the OGC formulating an across-the-board pilot property benchmarking scheme in 2005. This aimed to provide good quality and consistent real estate data in an effort to help inform decision making across the entire central government estate. It is designed to:

- put forward a standardized measurement framework;
- define efficiency (quantitative) and effectiveness (qualitative) in a consistent manner for all departments to reference;
- test the performance of 130 buildings in four departments;
- build a consistent, cross departmental database to track property performance over time; compare it with other departments and the IPD dataset;
- deliver reliable management reports.

The stated aims of the pilot benchmarking exercise are to create a system that will aid central government to more effectively manage its estate by highlighting:

- high cost buildings suitable for disposal;
- buildings that need urgent review and action on their inefficiencies and ineffectiveness;
- buildings with low office densities which could be suitable for intensification, rationalization and/or reconfiguration.

The system will use an electronic property information mapping service (e-PIMS) which pulls together spatial, utilization, ownership (lessee) and lease information, as shown in Figure 5.1.

Users are able to locate the property on an electronic map and instantly access all data applying to it. The system enables users to search for overused, underutilized and vacant space within offices, utilizing the benchmarking data to inform strategic decisions and pinpoint problem areas. The system is mandatory for all departments.

The performance framework is made up of three levels in a hierarchy and the analysis is split into two major components of Efficiency and Effectiveness.

LEVEL A: these are the overall headline numbers

example: for the overall Efficiency this is measured by the total occupancy costs per full time equivalent employee

LEVEL B: these are the key performance indicators (KPIs)

example: the occupancy costs per square metre

LEVEL C: these are the detailed measurements which contribute to the KPIs

example: amount of downtime in the office due to facilities failure

Each indicator is scored with a number, e.g. 109, this is a measure of how the indicator performs in comparison to the IPD Occupiers dataset thereby measuring against good practice and private sector standards. The scoring is percentage based, a score of 100 means that the indicator in question is performing as well as the

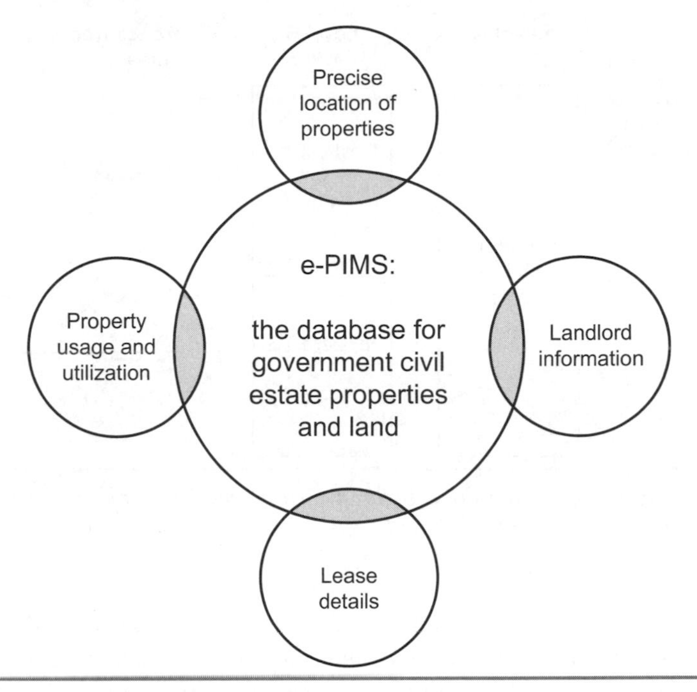

Figure 5.1 The e-PIMS database structure

benchmark (100%). A score of over 100 means that the indicator is performing better than the benchmark and below 100 the indicator is performing worse. Any indicator performing over or under 10% is highlighted to emphasize the best and worst performing assets as measured in comparison to the OPD benchmark.

The key benefit of this hierarchical system is that each component of the framework can be broken down into individual key performance indicators, or sub-components, to enable a clear and detailed identification of where good or weak performance exists within a portfolio.

Figure 5.2 shows a hypothetical set of outcomes for a building compared to the IPD Occupiers dataset and is used to give an impression of the framework only.

In the example above the EFFICIENCY indicators are demonstrating that a cost per FTE of £3,677 gives this office a score of 148 for overall efficiency when compared to the OPD dataset. In other words it is performing 48% better than expected for this Level A category indicator of efficiency. Looking at Level B in the hierarchy it shows that both KPIs are performing above expectations which have contributed to the Level A performance. The cost per square metre is 16% better than expected and the space per FTE is 42% better than expected. These differences can also be shown as absolute differences from the mean; in this case the office is costing £69,000 less than the benchmark and is accommodating 42 more FTEs.

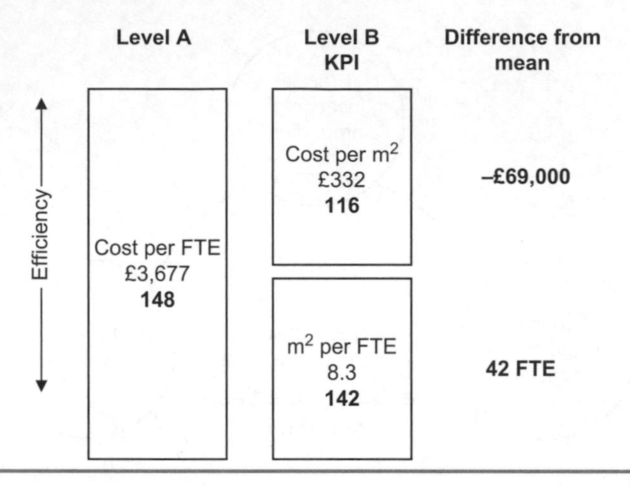

Figure 5.2 An example from OGC *Better measurement, better management* report (2006). *(© Crown Copyright)*

Beneath these headline results the detailed outcomes of Levels A, B and C will provide additional information that will allow managers to examine specific targets, for example environmental performance and target properties that fail to match the benchmark. They will then be able to find out which of the environmental indicators is the problem and begin to formulate strategies for improvement. A detailed illustration of the performance framework is illustrated in Figure 5.3.

The aggregated data can of course be used to look at the whole portfolio to find distributions and deviations from set targets and to compare regions. More examples of how this information can be used can be found at www.ogc.gov.uk.

We believe that the framework approach is logical and effective and provides useful data and consistent and meaningful comparisons with the private sector. It also provides a framework that can be adopted by the private sector. One deficiency that we note is that, unlike the models used by the United States General Services Administration (GSA), it does not appear to immediately offer the opportunities to use the framework as a scenario building tool to investigate the adoption of contemporary modes of working and innovation, which impacts upon the need for space.

The GSA cost per person model is an Excel based tool designed to enable users to benchmark their occupancy costs. Unlike the OGC/IPD model it is freely available from www.gsa.gov/cppm, is highly transparent and provides a tool for *'what if'* scenarios. The GSA Office of Real Estate cost per person model is one GSA's seven original government-wide performance indicators that continues to be seen as high importance. The model estimates the average cost per person in each of the following areas: real estate (space usage); telecommunications; information technology and alternative work environments. We like the GSA model because it promotes the

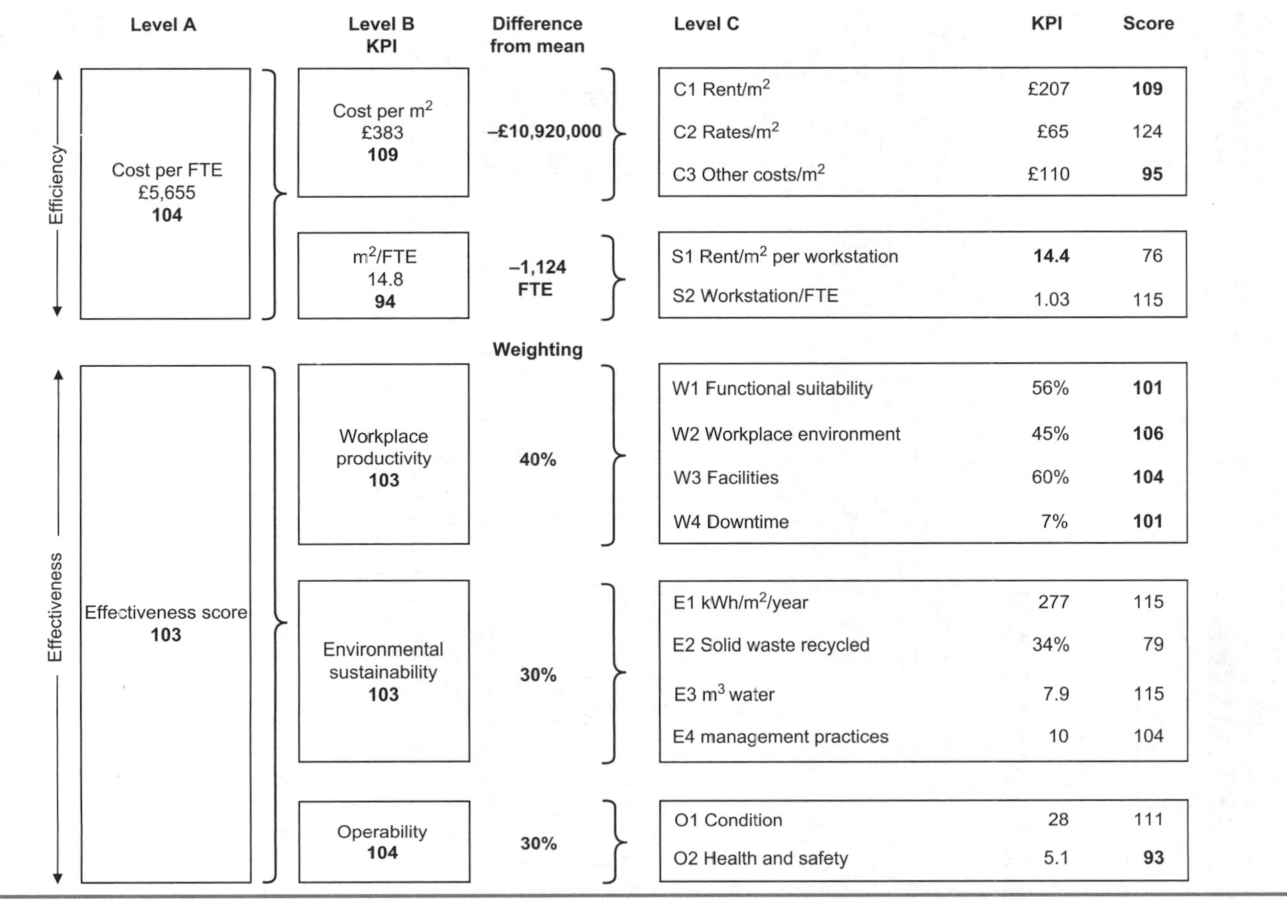

Figure 5.3 The OGC Performance Framework in detail: an example from *Better measurement, better performance* (2006). *(© Crown Copyright)*

SCENARIO TOOL To begin, click the RESET button. RESET

IMPORTANT: You must complete the *Cost per Person Model* in full before using the Scenario Tool.

POTENTIAL COST SAVINGS:

CLICK HERE TO VIEW POTENTIAL COST SAVINGS GRAPHICALLY

STEP 1	STEP 2	STEP 3
The Employee to Workstation ratio drives potential real estate cost savings. A higher ratio reduces the number of workstations, implying that more employees will telecommute or desk share. A lower ratio means more employees require workstations full time. Input a new Employee to Workstation ratio and press the TAB key. Your potential real estate cost savings will appear below.		
Employee to Workstation Ratio	Workstations	Teleworkers

Number of Employees:

	New Ratio	Current Ratio
Employees		
Workstations	1.0	

Number of Workstations

	New	Current
full time		
+		
shared		
=		
TOTAL		

No. of teleworkers working at:

	New	Current
telecentres		
home		
TOTAL	0	
No. days/week at telecentre		

Figure 5.4 The GSA Scenario Tool

analysis of alternative working patterns including hotelling and desk sharing, and allows users to model the cost saving impacts of introducing innovation. Whilst the model is simplistic and should be used with caution we do believe that the ability to plan alternative strategies and observe the impact on cost is beneficial.

The Scenario Tool input screen is shown in Figure 5.4. This in turn produces graphs that demonstrate cost savings for each scenario examined. We believe this approach is useful because it links to contemporary flexible working practices, the concept of working without walls and the preferences of the next generation workforce (Generation Y), as discussed in Chapters 1 and 6.

Benchmarking in Action 3

The IPD Occupiers International Total Occupancy Cost Code

The first Total Occupancy Cost Code (TOCC) was launched by IPD in 1999. It provided a comprehensively defined metric for capturing all occupancy costs involving rent, tax, fit out, furniture, building operation, business support and management, and according to IPD came to be used increasingly in the UK as an industry standard for measuring costs. The International Total Occupancy Cost Code (ITOCC) was launched in 2001, reflecting feedback and experience from users of the original Code. IPD justify the need for the cost code in terms of real estate asset managers requiring information to:

- make decisions better and quicker in a rapidly changing world;
- monitor performance in estates and facilities;
- communicate effectiveness and value-added to the rest of the organization.

They state that keeping accommodation costs under control is the first priority of most real estate/FM managers. An important feature of the ITOCC is the standard terminology, consistent definitions and standardized depreciation provisions which prevent errors when comparing assets as discussed in the 'apples and pears' section above.

The ITOCC is divided into five major categories, labelled A to E:

Heading A: Real estate occupation costs
Heading B: Adaptation and equipment costs
Heading C: Building operation data
Heading D: Business support costs
Heading E: Management costs for real estate and facilities

Within each of these major categories are a series of sub-categories, for example 'A4: local property taxes' and 'B2: furniture and equipment'. This allows users to drill down into detailed cost information for each property and compare it against their own portfolio and averages for all users or specific sub-sets of users.

An important part of the Code is the inclusion of both Revenue (annually occurring) items – such as rent, service charge and property taxes and Capital (one-off) costs such as telephone equipment, repairs and furniture.

The ESSENTIAL components of the ITOCC are therefore set out as:

	REVENUE ITEMS (X)	CAPITAL ITEMS (Y)	TOTALS (Z)
A Occupation	AX Rent, local property taxes	AY Acquisition costs	AZ = AX + AY
B Adaptation	BX Leasing costs	BY Fit out furniture	BZ = BX + BY
C Operation	CX Maintenance, cleaning	CY Major repair item	CZ = CX + CY
D Support	DX Reception, post room	DY Telephone system	DZ = DX + DY
E Management	EX Fees/internal	EY Major management expenses	EZ = EX + EY

The Total Occupacy Costs = TX + TY *where*:
Total Revenue Costs (TX) = AX + BX + CX + DX + EX
Total Capital Costs (TY) = AY + BY + CY + DY + EY

As mentioned at the start of this chapter the code has uniform depreciation timescales for each of its components to ensure consistency, some examples are given below:

B1 Fit out and improvement 7 years
B2 Furniture and equipment 5 years
C3 Internal repair and maintenance 5 years
C5 External and structural repair and maintenance 10 years
D1 Telephones 3 years
D6 Reprographics and printing 3 years

The documentation supporting the ITOCC is comprehensive and very clear, providing detailed explanations of each category within the costs code, helping to ensure consistency of approach and data submitted to IPD for inclusion in the benchmark. Two examples of the guidance are set out below.

As can be seen the Code provides detailed guidance on each element to ensure consistency and robust and reliable data collection. Readers are recommended to consult the full ITOCC Guide which is currently downloadable on request from the IPD website (www.ipdoccupiers.com).

ITOCC: Guidance Examples

A4 Local property taxes:	The full annual real estate, occupational and environmental tax liability arising under national and local laws and regulations within the subject country, state and municipality. All rebates should be averaged out across the period to which the rebate relates.
Includes:	any taxes or rates arising directly from the occupation of real estate and levied on the building itself or upon the occupiers of the building. Environmental taxes and charges for parking should be included.
Excludes:	all business and sales taxes that are levied on business profits and sales as distinct from the occupation of the building.
B2 Furniture and equipment	The total annual cost of all furniture and equipment relating to office, sales, production, storage, reception, workstation, meeting, training, boardroom and kitchen areas.
Includes:	desks, chairs, tables, soft furnishings, lighting units, pedestals, screens, curtains, drapes, blinds, shelving, cambers, storage cabinets, video conferencing equipment, works of art, storage bins, fire extinguishers, audio and video conferencing equipment, mechanical handling equipment, waste compactors and audio-visual equipment.

Recommended write-off period = 5 years.

KEY OCCUPANCY COST RATIOS

The code documentation provides a useful list of what they call Key Occupancy Cost Ratios (KPIs) categorized into business-linked ratios; occupancy cost ratios; and utilization ratios. These can all be measured and used as benchmarks within the ITOCC methodology:

Business-linked Ratios

Measures Relating Total Occupancy Costs to Wider Key Business Measures

Total Occupancy Costs per unit of revenue.
Total Occupancy Costs divided by total operating costs (for an organization, business unit or building).
Total Occupancy Costs per unit of business output (according to the business, e.g. units manufactured, insurance policies processed etc.).
Total Occupancy Costs per member of staff.
Change in relation to all these indicators over time.

Occupancy Cost Ratios

Measures to Relate and Illustrate Occupancy Costs

Total Occupancy Costs per square metre/per workstation/building unit.
Individual category costs per square metre/per workstation/building unit or % share of total.
Total Occupancy Costs of vacant or sub-let building divided by the total costs of all real estate/facilities.
Change in relation to all these indicators over time.

Utilization Ratios

Measures to Relate and Illustrate Space Utilization

Occupied space per worker/workstation.
Net : gross internal area ratio.
Ratio occupied, vacant and sub-let as % of floor space total.
Average visit time (building users divided by building user hours).
Building users per year.

The Code is comprehensive and flexible and can be and is used as a strategic benchmarking tool. It may, however, be expensive to administer and very demanding in terms of the data capture requirements, for smaller firms, especially in the UK, the

Total Office Cost Survey provided by Actium Consult (see Case Study 2) may be more appropriate.

Benchmarking in Action 4

Jones Lang LaSalle and the OSCAR Service Charge Benchmark

OSCAR was first published in 1983 and the latest edition (2007) available at the time of writing (2009) is the 24th in the annual series. The highly respected OSCAR work is regarded by the RICS as a considerable contribution towards their objective of more consistency and transparency within the property management sector. The RICS believes that as so many managers are adopting it as the industry standard for cost classification the size of the sample in OSCAR will continue to grow, making this data even more relevant and robust.

The dataset is drawn from audited or certified Service Charge accounts with 'year end' dates falling within the previous calendar year. The database is split into two property types, air-conditioned and non air-conditioned. In accordance with good benchmarking practice the system uses clearly defined and transparent terms to ensure an 'apples and apples' approach. Examples of these definitions, collected from the OSCAR website are:

- in line with market practice, OSCAR data are quoted excluding VAT;
- in order to allow level benchmarking every effort has been made to exclude Exceptional Expenditure from the analysis;
- metric equivalents have been shown where practical, rounded to the nearest convenient equivalent, but the analysis is principally in imperial data;
- the costs do not include the direct recharges borne by occupiers in relation to their own demised areas;
- interest payments credited to the service charge accounts are currently excluded from the dataset to avoid potentially distorting the underlying operational costs;
- in-line with market practice, any insurance costs are removed from OSCAR.

OSCAR provides a very valuable benchmark in allowing users to examine their own service charge costs either as a total or broken down into categories for Retail, Office and Retail Park premises in the UK and Frankfurt. The geographical coverage may grow in the future, giving a further international dimension to the system.

Figure 5.5 sets out three extracts from the 2007 OSCAR that give an indication of the benchmarking information available (available upon registration at www.oscar.joneslanglasalle.co.uk).

Holistic Real Estate Benchmarking in Practice

The following example is a simplified case study based on a real organization but anonymized and simplified for the purpose of this illustration.

A food manufacturing company has grown organically through acquisitions of other businesses and has production, warehousing and office facilities scattered across England, Scotland and Scandinavia. It has ambitions to float on the stock market but recognizes the fact that its real estate holdings are haphazard and inefficient and have not been integrated into its corporate strategy, which has largely been focused on aggressive growth and expansion.

Extract One

2007 Office OSCAR: Total service charge by location

	Central London	Other locations
A/C	£7.80 per sq ft	£5.60 per sq ft
Non A/C	£6.90 per sq ft	£4.60 per sq ft

Extract Two

2007 Office OSCAR (Air-Conditioned Buildings): Ranking of component costs within the service charge, in descending order of relative spend

Rank	Component	2007 %	2005 rank and %
1st	M&E Services	19.97%	1st 20.62%
2nd	Security	18.95%	2nd 18.82%
3rd	Electricity	12.97%	3rd 12.66%
4th	Cleaning & Environmental	11.22%	4th 11.49%
5th	Site Management Resources	9.48%	5th 9.79%
6th	Fabric Repairs & Maintenance	9.18%	7th 8.72%
7th	Management	8.31%	6th 9.03%
8th	Gas	4.37%	8th 3.78%
9th	Lifts & Escalators	3.94%	9th 3.60%
10th	Water	1.60%	10th 1.48%

Figure 5.5 Three extracts from the OSCAR benchmarking service (2007). It can be seen that the OSCAR service provides occupiers with a valuable set of benchmarking data against which they can compare their own service charge costs. The system provides detailed analysis so that occupiers can observe if a particular element is significantly at variance with locational averages and monitor volatility on a year by year basis

Extract Three

2007 Retail OSCAR: Average service charge components by percentage

	Enclosed Air-Conditioned	Enclosed Non Air-Conditioned	Part Enclosed	Open	Average
	2007 (2006)	2007 (2006)	2007 (2006)	2007 (2006)	2007 (2006)
Fees	7.54% (–)	8.20% (–)	9.12% (–)	9.54% (–)	8.60% (–)
Site Management	15.83% (–)	17.70% (–)	25.53% (–)	21.12% (–)	20.05% (–)
Management Total	23.42% (21.69%)	24.41% (21.67%)	23.12% (25.12%)	30.09% (30.22%)	25.26% (24.70%)
Utilities	9.89% (7.49%)	7.34% (6.05%)	5.78% (7.34%)	2.92% (3.78%)	6.48% (6.14%)
Soft Services Cleaning	17.42% (–)	16.47% (–)	17.85% (–)	18.81% (–)	17.63%
Soft Services Marketing	7.94% (–)	7.64% (–)	6.53% (–)	13.75% (–)	8.97%
Soft Services Security	18.78% (–)	18.49% (–)	16.13% (–)	23.12% (–)	19.13%
Soft Services Total	44.14% (48.10%)	42.59% (47.88%)	40.50% (43.06%)	55.68% (52.55%)	45.73% (47.90%)
Hard Services	19.88% (20.28%)	24.46% (23.13%)	15.01% (23.77%)	10.22% (12.07%)	17.39% (19.81%)
Insurance	0.90% (0.99%)	0.55% (0.87%)	2.00% (0.72%)	0.08% (0.60%)	0.88% (0.80%)
Exceptional Expenditure	1.76% (1.050%)	0.65% (0.40%)	13.59% (0.00%)	1.01% (0.73%)	4.25% (0.55%)

Figure 5.5 — Cont'd

Real estate and business consultants were instructed to review the portfolio and business operations and prepare recommendations for alignment of the real estate portfolio with the business strategy and to identify opportunities and threats within the portfolio.

The consultants visited each property and undertook a SWOT analysis of each real estate asset (see Chapter 1). They then prepared a strategic plan to analyse the portfolio and business units and identified a set of measurements and internal benchmarks they wished to observe, measure or calculate. These were classified under the following headings: Business, Real Estate, Financial and Hybrid. The performance measurement calculated under each heading are set out in Table 5.3.

The results of the review of the portfolio, the SWOT analysis, the calculation of the internal performance comparing the business units against each other, plus the external benchmarking revealed significant threats and opportunities within the

TABLE 5.3 Performance Measurement and Benchmarking Undertaken by the Food Manufacturing Company

	Internal performance measurement	External benchmark
Business		
	ROCE	Other similar quoted companies published accounts
	Gearing	
	Liquidity ratio	
Real estate		
	Leasehold to freehold ratio	Donaldsons–Lasfer Research
	Total occupancy costs per square metre	Actium Consult
	Total occupancy costs per person	Actium Consult
	Space analysis: space per person in the office parts of the portfolio	OPD
	Site utilization and density	
	Space utilization – offices hours per working week by individual	
	Space utilization – meeting room use per day	
Financial		
	Market value to book value to alternative use value	
	Freehold property returns to business investment returns	IPD and company accounts
Hybrid		
	Distribution distances – food miles of products travelled	
	Occupancy costs as a percentage of operational costs	
	Profit per square metre for each business unit	

portfolio and identified strategies for adding significant value to the business. For example:

- opportunities for cash generation through sale-and-leasebacks which reduced the high volume of freehold ownership;
- rationalization of a poorly performing business unit that was in a remote location, which extended both the distribution costs and food miles; the portfolio review discovered an alternative use value (residential) significantly in excess of the book value;
- opportunities for rationalization through site intensification and underutilized density;
- opportunity to relocate head office to alternative site, generating cash and reducing space per person;
- opportunity to introduce hot desking and contemporary ways of working in new office location to improve productivity;
- identifying weak production per square metre, compared to both internal business units and external competitors, identifying opportunities for improvement;
- reducing meeting rooms which were very badly utilized and using flexible offices (MWB) on a 'pay as you go' basis.

The opportunities for strategic change were discovered and unlocked through the benchmarking process and provided evidence of both underperformance and good practice. The example illustrates how important it is to integrate both business and real estate performance and benchmarks together and to engage a wide range of stakeholders. These should include the company accountant, business managers, asset managers and facilities managers, plus in this case the Human Resources department to ensure the new working practices are embedded within the strategic asset plans.

Note how this example resonates with the added value approach and the framework presented by Lindholm and Leväinen (2006) to increase shareholder value through:

- increasing the value of assets;
- promoting market and sales;
- increasing innovation;
- increasing employee satisfaction;
- increasing productivity;
- increasing flexibility;
- reducing costs.

Benchmarking and CREAM Strategy

The above example demonstrates the powerful outcomes that can be achieved when benchmarking is integrated with corporate and real estate strategy. However, we end this chapter with a word of caution. We have seen that benchmarking is a powerful tool that is being rolled out across UK central government and is being promoted by IPD Occupiers for corporate occupiers as a strategic tool. However, we return to the dangers of using benchmarking as a blunt driver for strategic action. The use, for example, of a space standard, say the 12 square metre maximum prescribed in the OGC approach, does not necessarily engage with the complexity of office design, layout and utilization and the innovative ways in which space is being used by organizations. Fundamentally it tends to assume territorial space and ignores the main tenet of new ways of working, i.e. the user has autonomy to use a variety of spaces that suit the work process not being a slave to a static workstation or work setting.

Osgood (2002b) defines a more appropriate approach in his Strategy Alignment Benchmarking Component, which is comprised of a three-step approach that is designed to take people beyond generic to organization-specific best practices and from one- to multi-dimensional analysis. Osgood argues that

> *benchmarking should be used to achieve competitive advantage rather than to simply copy and catch up.*

He warns of the dangers of universally applying benchmarks that have achieved 'industry status'. He uses the example of space and space planning and warns, as we would also, of the need to incorporate a consideration of work styles and practices alongside the one-size-fits-all space allocation. So in his model, the IPD Occupier and other models described above would be part of the stage 1 evaluation strategy set out in Figure 5.6 to inform overall space standards and occupancy costs.

The user would also have to examine a continuum of appropriate space solutions appropriate to different operations and recognize that each of these will demand a different space allocation. Benchmarking in its wider sense as a process examining other institutions could be used in stage 2 to help inform innovation in workplace settings, use of technology, and working without walls, as discussed in Chapter 6.

In stage 3 an informed consideration of space types and work styles could be used and a new set of space standards developed for each setting. This is described by Osgood (2002b) as benchmarking group workflow and examines the types of workflow, the information or knowledge flow, and the individual or group collaboration. This is set out in Figure 5.7.

This approach allows planning of an organization's facilities to work within a hierarchy of benchmarks, appropriate for each operation and to test different configurations against space standards and work practices. It recognizes the complexities and subtleties of designing in productivity (as discussed in Chapter 6) and we would argue that it requires a much more sophisticated and informed

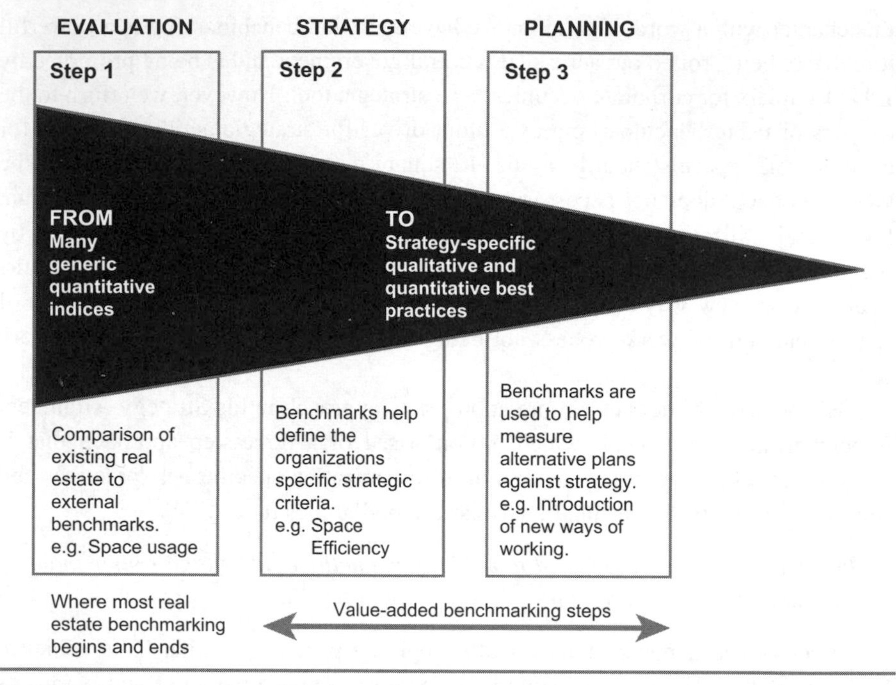

Figure 5.6 Strategic benchmarking alignment model. *(From Osgood, 2002b; reproduced by permission of* Site Selection *magazine)*

approach to benchmarking than simple comparisons against averages or expected industry standards. It also demonstrates how a dual approach is required:

- external benchmarks collected and analysed by an independent organization to set the context;
  plus
- internal generation and testing of strategic options appropriate for the organization; informed by best practice and innovative early adopters of new practices, in a benchmarking process.

The Balanced Scorecard

We wish to mention briefly the importance of balance in benchmarking and the Balanced Scorecard as developed by Kaplan & Norton (1996). The Balanced Scorecard is a management tool that maps an organization's strategic objectives into performance metrics and targets organized into four sections: Financial Performance, Customers, Internal Processes, and Learning and Growth. Discussion of the

Types of workflow	Information exchange	Emphasis Individual concentration *vs.* Group collaboration	Allocation required
❶ No relationship There is administrative linkage but no formal, regular communication between people in the group	**Occasional to frequent exchange of explicit information**	✓	**3%**
❷ Technology Technology serves as the connection between people in the group		✓	**7%**
❸ Formal independent Technology and planned, infrequent, face-to-face interaction with people in the group	**Frequent exchange of explicit and occasionally tacit information**	✓	**11%**
❹ Independent Occasional, primarily planned, infrequent, face-to-face interaction with people in the group		✓	**29%**
❺ Formal individual contributor Frequent, primarily planned, face-to-face interaction with people in the group		✓	**8%**
❻ Individual contributor Frequent, planned and impromptu face-to-face interaction with people in the group		✓	**20%**
❼ Team contributor Frequent, impromptu face-to-face interactions between most or all members in the group	**Frequent on-going exchange of tacit information**	✓	**6%**
❽ Team Tasks and problems are shared by the entire group in an on-going series of impromptu face-to-face interactions		✓	**16%**

Figure 5.7 Identifying workflow and practices for detailed strategic benchmarking of space. *(After Osgood, 2002b)*

The Financial Perspective	The Business Growth Perspective
Financial result Return on assets Value of properties (book and/or market values) Amount of sold real estate Vacancy rate Leased *vs*. owned real estate	Vacancy rate Space per employee Occupancy costs Operation costs Transaction costs Return on assets Sustainability achievements, recycling, energy consumption etc. Budgeting, targets, scheduling of projects Business unit's satisfaction with CREM performance
The Learning and Growth Perspective	**The Customer Perspective**
Professionalism and expertise of CREM staff Competence of the staff Customer service Age profile Education profile Reward and motivation systems Atmosphere of the environment Training and development days Absenteeism Turnaround Research and development capacity	Internal customer satisfaction (of the workplace) External customer satisfaction of the workplace Cost–quality ratio Customer satisfaction of services Cost–quality ratio of facilities services Satisfaction for CREAM department (delivery of facilities services, reporting, response etc.)

Figure 5.8 CREAM benchmarks in a Balanced Scorecard framework. *(Based on Lindholm & Nenonen, 2006)*

tool in detail is beyond the scope of this book, but we set out in Figure 5.8 a framework of metrics in the Balanced Scorecard format based upon that developed by Lindholm & Nenonen (2006).

In Case Study 1: Bruntwood we examine their approach to customer service benchmarking and confirm that they use the Balanced Scorecard to good effect, linking, for example, their customer service metrics to training of their staff, appraisal and reward, which impacts upon customer retention and financial performance.

Conclusion

In this chapter we have evaluated a number of benchmarks and suggested the need for a combined perspective of corporate real estate. As Liow & Nappi-Choulet (2008) comment:

> *A combined framework helps to position the strategic role of CREAM in the context of the 'whole firm' that reflects the integration of trading and real estate activities.*

Performance measurement and benchmarks can, as with our example, unlock a deep understanding of the opportunities, risks and threats within a business and real estate portfolio and inform strategic options for improving the business performance through real estate based interventions.

We have also made explicit linkages to the way in which benchmarking can help to identify opportunities for adding value. This approach is more likely to generate strategic options tailored to an organization's business context, culture, aspirations and performance targets, than a simple metrics-only approach.

Summary Checklist

1. Does your organization undertake systematic performance measurement and/or benchmarking. If not, why not?
2. Does your organization examine year-on-year performance and compare facilities internally, for example, total occupancy costs or space per person at each location? If not, consider starting a systematic measurement process.
3. Do you have an 'apples and pears' problem? Have you fully considered issues of compatibility, including:
   a. Measurement
   b. Type of occupation
   c. Periods of depreciation
4. Are you aware of the variety of benchmark clubs or publications available to externally validate your internal analysis? Consider which are the most suitable for your organization, in terms of cost, scale, time, relevance and data collection commitments.
5. Consider linking your company strategy and real estate strategy by using appropriate 'hybrid' indicators, for example:
   a. real estate costs as a percentage of production costs per business unit.
6. Consider a strategic approach as advocated by Osgood, rather than simplistic measurement of space. Are there strategic opportunities to align business goals, culture and human resource issues with space (see Chapters 6 and 7) which are revealed by the performance measurement/benchmarking process? For example, does underutilization of desks for some staff demand the consideration of hot desking or hotelling solutions. Or does the underutilization of meeting rooms suggest that on relocation a contemporary landlord (see Chapter 3) who offers meetings rooms on a 'pay as you go' basis make business sense?

7. Are your service charge payments too expensive ? Use the freely available OSCAR service to benchmark your costs against the average in the UK and Frankfurt (for Offices).
8. Analyse your occupancy costs using either OPD or Actium Consult Benchmarks to determine whether your occupancy costs are out of line with your competitors and examine which components of your occupation are excessive or performing well.
9. Consider a strategic approach to space usage and measurement as advocated by Osgood and consider the types of workflow and use of the space before applying an analysis of space per person.
10. Examine the balanced scorecard approach and consider the real estate metrics that should be included to support the business strategy of your organization.

References

Avis, M., & Gibson, V. (1995). Real Estate Resource Management: A Study of Major Occupiers in the UK. Wallingford, GTI: The University of Reading, Oxford Brookes University.

Bon, R., Gibson, V., & Luck, R. (2000). Annual CREMRU-JCI survey of corporate real estate practices in Europe and North America: 1993-2001. *Facilities, 20*(11/12), 357–373.

de Jonge, H., de Vries, J., & van der Voordt, T. (2008). Impact of real estate interventions on organisational performance. *Journal of Corporate Real Estate, 10*(3), 208–223.

Johnson, G., & Scholes, K. (1999). *Exploring corporate strategy text and cases* (5th ed.). Harlow: Prentice–Hall.

Jones, Lang LaSalle, in association with AEEBC. (2005). Leased Space Comparison in EMEA.

Kaplan, R., & Norton, D. (1996). *The Balanced Scorecard: Translating strategy into action.* Cambridge, MA: Harvard Business Press.

Lindholm, A., & Leväinen, K. (2006). A framework for identifying and measuring value added by corporate real estate. *Journal of Corporate Real Estate, 8*(1), 38–46.

Lindholm, A., & Nenonen, S. (2006). A conceptual framework of CREM performance measurement tools. *Journal of Corporate Real Estate, 8*(1), 109–119.

Liow, K., & Nappi-Choulet, I. A combined perspective of corporate real estate. *Journal of Corporate Real Estate, 10*(1), 54–67

Maire, J. (2002). A model of characterization of the performance for a process of benchmarking. *Benchmarking: An International Journal, 9*(5), 506–520.

Nourse, H. (1994). Measuring business real property performance. *Journal of Real Estate Research, 9*(4), 431–444.

OGC. (2006). Better measurement, better management: Effective management of the Government estate. http://www.ogc.gov.uk/documents/cp0134.pdf. Accessed 12.10.09.

Osgood, T. (2002a). The strategy alignment model: defining real estate strategies in the context of organizational outcomes. *Site Selection.* May 2002.

Osgood, T. (2002b). The strategy alignment model: benchmarking for competitive advantage. *Site Selection.* May 2002.

RICS Property Measurement Group. (2007). *Code of measuring practice: A guide for property professionals* (6th ed.). London: RICS Books.

Zairi, M. (1998). *Benchmarking for best practice: Continuous learning through sustainable innovation.* Oxford: Butterworth–Heinemann.

Chapter 6

Workplace Transformation

Introduction

Office environments have evolved over the past century in response to the changing nature of work. There has been a move from individuals working in cellular offices to teams of people working in open plan environments. Today's workplace allows people to interact freely, thereby supporting collaborative working. However, obtaining the right balance of interaction in an office environment without it turning into distraction requires a detailed understanding of the work processes undertaken in the office.

In this chapter we develop the context for the changing nature of office working. Work is no longer restricted to the office environment and issues relating to work undertaken out of the office and virtual working will be explored. Finally, we explore the linkage between productivity and the office environment with special focus placed on the role of the behavioural environment.

Evolution of the Workplace

To help set the context for understanding the modern office environment it is worth reflecting on how the office has evolved over the past 100 years. To truly appreciate the changing nature of the office environment it is important to establish the impact of the prevailing management style adopted by office managers.

In the early years of the 20th century Frederick Winslow Taylor developed a management style which was termed 'scientific management'. The methodology proposed was that by adopting time and motion studies the optimum way of performing a task could be identified. This led to the belief that work could be segmented into sub-components. This systematic and mechanistic view of work design saw the human element as a variable that needed to be designed out of the system. This approach meant that workers were given clear instructions which had to be followed precisely. One of the criticisms of the Taylorist approach was that it was

Corporate Real Estate Asset Management. ISBN: 978-0-7282-0573-4

deemed to dehumanize work. Since workers only worked on one part of a production line they never saw how the full production process worked; it was management's responsibility to monitor overall performance. This is a good example of 'command and control' management.

The scientific management principles developed by Taylor were specifically designed for factory-type activities. Frederick Taylor was aware of the limitations of these techniques, indicating that they would only work for employees who were happy to undertake routine, repetitive tasks. A good application of the scientific management principles was the Henry Ford Model T production line.

In 1904, the Larkin mail order headquarters, which was designed by the architect Frank Lloyd Wright, was opened in New York. Scientific management principles were transferred from the factory environment and applied to the office environment of the Larkin building. Based on the principle that the office workers would undertake process work, the building was designed in an ordered structure to ensure that the process working was as efficient as possible. This drive for process control meant the clerks were perceived as production operatives and given architect-designed desks, which allowed little freedom of movement. It could be claimed that this was the first example of the Taylorist office.

During 1924 and 1932 a number of studies took place at the Hawthorne plant near Chicago, which was part of the Western Electric Company. These studies are commonly known as the 'Hawthorne experiments'. The research directors, Elton Mayo and F.J. Roethlisberger, wanted to establish the link between the working environment and productivity. The starting research paradigm was, similar to the scientific principles of Frederick Taylor, that the work environment could be controlled and specific outputs, such as productivity, could be predicted (Roethlisberger & Dickson, 1939). Over the years two studies produced unexpected and revealing results about how people behaved in their working environment. The two studies were the illumination experiments and the bank wiring observation room study.

The first study aimed to link productivity levels with the lighting levels of the workers. The study was designed as a traditional experiment, with the view that changing the independent variable (lighting level) would have an impact on the dependent variable (productivity level). Early indications appeared to support the theory that lighting levels could have a causal impact on productivity levels. However, the researchers established that productivity levels continued to increase even when lighting levels were not increased. This led the researchers to propose that there was another intervening variable that they had not considered. Eventually, the conclusion reached by the researchers was that the workers were responding to the presence of the researchers and the fact that someone was showing an interest in their working environment. This effect is now commonly known as the 'Hawthorne effect'.

The second Hawthorne experiment observed workers in a bank wiring room. The pay received by the employees was determined by how much work they actually did, commonly known as piecework. The research team observed that higher productivity levels achieved by new employees eventually levelled out to be more in line with the experienced workers. This normalization process could be considered as a form of work restriction. Given that this research was undertaken during the time of the Great Depression it was an interesting conclusion that people were more willing to conform to the norm of a group rather than to receive individual financial reward.

The results of the Hawthorne experiments established that understanding employee productivity was far more complex than linking one variable to another. The social factors around a working environment are an integral part of understanding employee satisfaction and productivity. The identification of these social factors could be considered to be the start of the 'Human Relations' movement. The Hawthorne results were in contrast to the previous scientific principles of Frederick Taylor, which placed the manager as an independent observer of worker performance; rather, the Hawthorne studies identified that a manger was an integral part of the social system. The manager can impact directly on worker performance by motivating and communicating with the workforce. The Hawthorne experiments were a major step forward in understanding the relationship between people and their work environment.

Unfortunately, the lessons of the Hawthorne experiments were not applied to the office environment. Office designs in the 1950s adopted Taylor's scientific management principles, with individual cellular offices being reminiscent of individual production line workers. The traditional 1950s office building would be a narrow building with a long corridor down the centre of the building and cellular offices off to each side of the corridor. The individual cellular office acted as a cultural cue as to where you were in the organizational hierarchy. As people worked up the organizational ladder, they achieved larger office environments, clearly indicating their power and level within the organizational structure. The office layout encouraged individual private working and discouraged communication and team collaborative working. The kind of office layout could be considered as 'executive row' (Steele, 1983).

During the 1960s the open plan office environment began to appear. The creation of this kind of working environment addressed a number of issues that related to cellular office working; however, it also raised a number of potential disadvantages. A summary of the advantages and disadvantages of an open plan office can be seen in Table 6.1.

Burolandschaft was a specific type of open plan environment developed in Germany by the Schnelle brothers and introduced to the UK by Frank Duffy in the mid-1960s (Duffy, 1992). The Burolandschaft or 'office landscape' concepts aimed to address the weaknesses of the cellular and open plan office environments. The social factors identified by the Hawthorne experiments, such as group behaviour, were to be considered an integral part of the office design. Major consideration was given to how people interacted and Burolandschaft aimed to facilitate communication by association, with individuals and groups placed next to each other.

TABLE 6.1 Advantages and Disadvantages of Open Plan Offices

Organizational advantages	Individual disadvantages
Improved communication	Disturbance
Improved personal relations	Loss of confidentiality
Improved team spirit	Loss of identity and status
Greater equality	Loss of managers' privacy
Improved flexibility	Physical discomfort
Improved productivity	
Lower running costs	

During the early 1980s the concept of commons and caves was developed as a way of trying to establish the balance between open plan and cellular offices (Steele, 1983). Commons and caves proposes that instead of the office environment being either cellular or open plan, it could actually be both. The idea is that the office occupier can obtain the advantages of both open plan and cellular offices without the disadvantages. If an office worker wishes to work privately they can withdraw to a cave, and if they wish to interact with other colleagues they can by being in a common area. The physical application of the commons and caves concept is the Scandinavian combi office. A typical layout would be small cellular offices around the periphery of the office environment and an open plan area in the centre. People could stay in the private cellular offices to work individually and interact in the open plan area when they wished to collaborate or meet with other office colleagues. The commons and caves concept attempted to reconcile the connection between the workspace and the activities of the individuals using the space. Fritz Steele (1983) termed the relationships between individuals and their workspace as *'organizational ecology'*:

> *Organisational ecology is the area of reciprocal relations between people who work in organisations and their workplaces. It includes the impact of settings on individuals, groups, and the whole organisation, as well as the influence of organisational 'character' and dynamics on the design, management, and use of workplaces.*
>
> Steele, 1983, p. 65

The past 20 years have seen a significant impact on office design by the development of information communication technologies. It could be argued that the introduction of a personal computer (PC) onto everyone's desk has enabled increased electronic communication at the expense of face-to-face interactions. The physical positioning of a PC on an office worker's desk reinforces individual process work rather than group collaborative work. More recent technological developments including wireless technologies give the office worker a lot more freedom to work whenever and wherever they feel most appropriate. Some office workers are no longer constrained to their office desk facilitated by networked offices and ubiquitous computing.

Changing Nature of Work

The increasing challenges of globalization and competition mean that organizations have to be dynamic and agile if they wish to be competitive in today's business environment. In response to these challenges a range of management theories and strategies have been developed. Office work has had to evolve away from the static individual process working and towards more collaborative knowledge sharing. The shift away from individual process work to more knowledge working was identified by Peter Drucker in his 1959 book *Landmarks of tomorrow*, where he first coined the phrase *'knowledge worker'* (Drucker, 1959). Originally the term knowledge worker related to mainly professional workers. In today's society the term can be extended to include any kind of work that is involved in the creation and transfer of knowledge.

The acknowledgement that workers with knowledge were beneficial to the organization was counter to the Taylorist management philosophy, and more supportive of the human relations movement. In both *The age of unreason* (Handy, 1989) and *The empty raincoat* (Handy, 1994), Charles Handy argues that organizations tend to overlook their most valuable asset – their people. Handy's view was that human assets in an organization were becoming increasingly the core of the organization. Handy also questions the way that we organize our work life around the need to work 40 hours over five days. Since business cycles tended to increase and decrease depending on demand, an alternative approach would be a organizational structure that enabled more flexible working. Charles Handy called this approach the *'Shamrock Organization'*, because it had three components. The first component consists of the core staff; these are the main human assets within the organization. The second component consists of a more flexible workforce; these people have knowledge that the organization needs, but are used on a more flexible basis. The final component is the contractual fringe; these people are used during times of high business demand, but can easily be reduced if business demand falls. To ensure the shamrock organization works effectively there is a need for more team based structures.

The changing nature of work is a dynamic component which requires careful consideration when matched against property portfolio provision (Gibson, 2001). Gibson proposes that the three components of the shamrock organization can be matched with three corresponding property provisions. The core staff of an organization would be matched against any freehold property provision. The flexible staff employed on a cyclical basis could be matched against any property provision with short term leases. And finally, the casual contractual fringe workers could be matched against a 'pay as you go' property provision. This matching of work type with property provision allows an organization the flexibility and agility to adapt to changing business circumstances (Gibson, 2001).

The shift to a more knowledge economy has meant that repetitive process type working has either been replaced by computers or has been outsourced to a more cost-effective means of delivery, which may be in another country that has a lower cost base.

The move from individual routine processing work to more knowledge processing work changes the dynamics in an office environment. The changing dynamics means less solo private working and a shift towards more team based collaborative working.

Distributed Workplace

The changing nature of work now means that the office environment may not be the most appropriate place to undertake a specific work activity. The impact of information and communication technologies (ICT) and changing work processes means that work activities can be undertaken in a number of different places, and also in a number of different ways.

DEGW, the international architectural practice, has developed a model to represent the distributed workplace (Hardy et al., 2008). The distributed workplace model, as can be seen in Figure 6.1, can be used as a tool to help the creation and selection of the appropriate workplace strategies.

The distributed workplace model has three levels of privacy and accessibility, which are private, privileged and public. The work processes in each level can be undertaken in either a physical environment or a virtual environment. A description of these three levels of privacy and accessibility now follows (Hardy et al., 2008):

Virtual	⟵⟶	Physical
Knowledge systems e.g. virtual private networks/intranet	**Private** Protected access Individual or collaborative workspace	e.g. owned office, home office
Knowledge communities e.g. project extranets, video conference	**Privileged** Invited access Collaborative project and meeting space	e.g. clubs, airport lounges
Internet sites e.g. public chat rooms, information sources	**Public** Open access Informal interaction and workspace	e.g. cafes, hotel lobbies, airports

Figure 6.1 Distributed workplace model. *(From Hardy et al., 2008, p. 31)*

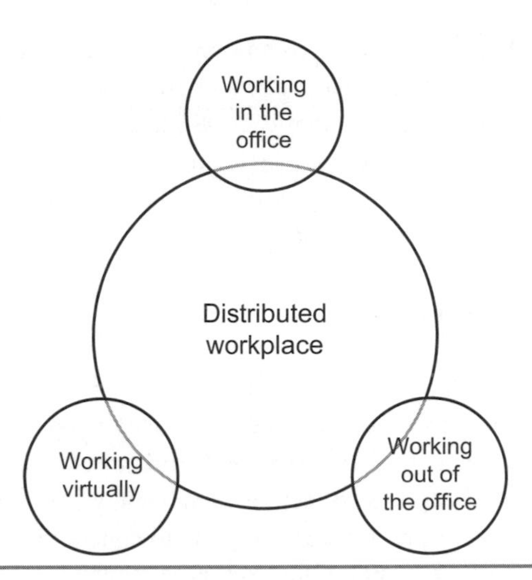

Figure 6.2 Components of the distributed workplace

- **Private.** This category entails either individual or collaborative workspace. This is a private space and has protected access. The physical environment that would support this type of working would be either the office environment or the home office environment. The virtual environment equivalent could be a protected intranet site belonging to an organization.
- **Privileged.** This entails collaborative working which could be undertaken in a meeting space within an organization that has invited access; this could be a project space. The virtual equivalent would be an Internet site that requires membership.
- **Public.** Within this category of working there is a requirement for informal interactions. These could take place outside the organization in places such as coffee shops. This type of interaction has open access and so Internet sites with public chat rooms could be the virtual equivalent.

The components of the distributed workplace can be considered to consist of a combination of the location of the work activity and the nature of the work activity undertaken. Figure 6.2 illustrates the three main components of a distributed workplace: working in the office, working outside the office and virtual working. We will explore each of the distributed workplace components identified in Figure 6.2 in greater depth in subsequent sections of this chapter.

Working in the Office

Whilst opportunities exist to work either virtually or outside the office environment, the office still has a role to play in today's business environment. The face-to-face interactions with either colleagues or clients cannot be underestimated. The modern office environment can provide meaning, a sense of belonging and can even be a place of experience (Jones Lang LaSalle, 2008). It can represent the culture and the branding of an organization as well as provide a variety of spaces to match a variety of work style needs. In this section we explore how office space can be provided to support different modes of office working.

Space for Concentration

In the modern office environment there is still a requirement to withdraw from office activities and find a private quiet place to concentrate on a piece of work. There are times in a day when concentration and focus are required, especially when undertaking activities such as thinking, reflecting, analysing, writing and problem solving (Gensler, 2008a). Giving office workers the time to concentrate on a particular task without distractions and interruptions can potentially enhance their productivity and performance.

Research undertaken by Gensler indicates that people in the United Kingdom spend an average of 59% of the time working on focus work (Gensler, 2008a). In contrast, people in the United States spend 48% of their time working on focus work (Gensler, 2008b).

The Welcome Workplace study, which is part of the Design for the 21st Century initiative, identifies the frustrations that office workers can experience when working in an open plan environment.

> *I find it hard to concentrate. My team can all be talking on the phone and I have to concentrate on a financial report. That was a challenge, and continues to be.*
>
> **Smith, 2008, p. 14**

Open plan environments should therefore incorporate private work areas that can allow individual focus working. These private work areas would represent the caves in the concept of commons and caves (Steele, 1983). To ensure the private work areas maintain their function, office protocols would have to be introduced to ensure optimum working.

Space for Collaboration

A key ingredient of office work is that collaboration can be undertaken with colleagues in the office environment. Space needs to be provided to allow work with another person, or a group of people, to be undertaken. Collaborative spaces can take the form of project spaces or team spaces. It is important that people feel they can converse freely and easily with their colleagues in this space without fear of distracting or interrupting other office workers. The activities undertaken in this collaborative space would include:

> *sharing knowledge and information, discussing, listening, co-creating, showing, brainstorming. Interactions may be face-to-face, by phone, video, or through virtual communication.*
>
> **Gensler, 2008a, p. 5**

Collaborative space may be space that could be allocated to a project for the duration of the project. This means that the space is owned by the team during the project period and can be deemed to be a permanent 'hub' (Smith, 2008). The collaborative space can give a project an identity. Once the project is complete, the collaborative space can be reorganized and reused for the next project and group. The spaces created for collaborative working could be considered to represent the commons in the concept of commons and caves (Steele, 1983).

According to the Gensler Workplace Survey 2008, companies in the United States spend an average 32% of their time collaborating (Gensler, 2008b), as compared to UK companies who spend on average 22% of their time collaborating (Gensler, 2008a). A report by Jones Lang LaSalle (2008) proposes a shift towards more collaborative working and states that progressive organizations are reducing individual space and increasing the amount of shared space to a balance of approximately 50/50.

Space for Contemplation

Office workers need spaces to withdraw to when they need time to rest and recuperate. In a results-driven business environment, it would be easy to work constantly throughout the day without even a lunch break. However, since the added value the office worker brings to the organization is their creativity, knowledge and motivation, these need to be preserved if the office occupier is to work at their optimum productivity. Spaces need to be provided which allow people to get away from the hustle and bustle of office life. This may include a rest room, a gym or even a Zen garden. The aim is to provide space that allows office workers to recharge their batteries.

The following quote from the Welcome Workplace study indicates the importance of being able to contemplate, and if space is not provided in the office then people will find the right kind of space elsewhere:

> *If you look around the coffee shops around the building, they are full of people reading and writing presentations and that sort of stuff.*
>
> **Smith, 2008, p. 18**

Space for Socialization

One of the benefits of working in an office environment is that it allows people to develop and maintain relationships with their work colleagues. This socialization could be classed as the social capital of an organization (Gensler, 2008a). Gensler defines social capital as:

> *Work interactions that create common bonds and values, collective identity, collegiality and productive relationships.*
> *Talking, laughing, networking, trust building, recognition, celebrating, interacting, mentoring, and enhancing relationships.*
>
> **Gensler, 2008a, p. 5**

Using the above definitions, the Gensler Workplace Survey 2008 established that in both the United Kingdom and in the United States only 6% of people spend their time in social activities. Since the benefits of social interaction can be improved productivity and performance of office workers, it makes sense to provide more social interactive space. This space could include small breakout areas, more club space that incorporates a range of different types of layout, or even a corporate cafe. Increasingly the corporate cafe is being used as a 'hub' within an organization for employees to socially interact and network, whilst at the same time offering the opportunity to maintain and develop relationships with other office colleagues.

Space for Learning

It could be argued that an essential requirement of a knowledge worker is the ability to acquire new knowledge or intellectual capital. This ability to learn can be formal learning achieved through education and training courses, or informal learning obtained by working with others and thereby developing experience and expertise.

Learning spaces could be individual spaces for concentration and contemplation. Alternatively, learning spaces could be group interactive spaces for collaboration and socialization.

The results from the Gensler workplace survey 2008 indicate this to be an area that needs to be developed, since only 6% of companies in the United States spend their time learning compared to only 4% of companies in the United Kingdom. Since office environments are increasingly becoming knowledge exchange and creation centres it would appear that integrating learning to develop intellectual capital is an essential ingredient of any workplace strategy.

Working Out of the Office

It is becoming increasingly acknowledged that the workplace cannot support all the different work styles that the new office worker undertakes. An integral part of a flexible working strategy is the supporting management culture. It is less important to monitor the time spent sat in an office at a desk and more important to monitor people's outputs. This shift in emphasis from inputs and 'presenteesim' to outputs and performance enables the employee to choose the most appropriate environment to support their current work needs.

Possible working environments outside the office:

- **Homeworking.** The blurring of work–life boundaries can have benefits to employees in that it gives them the flexibility and autonomy to work when and where they feel most productive. However, the other side of this flexibility is that the separation between work life and home life can tend to be hard to differentiate.
- **Serviced offices.** The office user has ultimate flexibility of office space, since space can be paid for on an hourly, daily, weekly or monthly rate. In addition to office space other facilities such as reception and administrative services could also be provided.
- **Client's premises.** By being in the physical presence of clients or customers, it is possible that a closer relationship could be established. The ability to develop a rapport with the client or customer provides a possible foundation for customer relationship management (CRM).
- **Cafes.** Places like Starbucks are becoming a surrogate to the office environment. They offer an opportunity for private reflective time or alternatively they can be places where meetings can be undertaken for a short duration of time. All this flexibility comes at the price of just a coffee.
- **Public spaces.** Areas such as libraries, local art centres, open public spaces and parks offer a range of different types of environments for individual contemplative work or more collaborative group working. Given the right weather conditions, *'laptops in the park'* could be a both enjoyable and productive working environment.
- **During transportation.** Walking along a carriage of an Intercity train one can usually see an array of laptop computers with screens showing reports, spread sheets and presentations. The tables on the train that accommodate four people become a meeting point, allowing both informal and formal interaction to take place. Another area that offers an opportunity for work is the airport lounge. This allows a more efficient use of time, since many hours can be spent waiting to board an aeroplane.

- **Hotels.** Most hotels now offer wireless Internet connections giving people the possibility of constantly being connected to an e-mail account. In addition the hotel lobby could be used for informal meetings with more formal meetings being undertaken in a booked room.

Working Virtually

Whilst face-to-face interactions with colleagues have great benefits, especially during the creative phase of collaboration, they are not always possible due to time pressures and the geographical positioning of the individuals involved. The developments in emerging technologies mean that most of the benefits of face-to-face interactions can now be achieved by using the right type of technology.

This type of working can be classified as virtual working, as interactions and collaborations with colleagues can still take place but just in a different form. The choice of virtual working method will be dependent on level and nature of interaction with colleagues and the level of response and urgency of the interaction.

The simplest level of virtual working could be the use of technology to send a piece of information to a colleague, such as an e-mail or text message. The information sent could be considered to be one-way communication, which means a response from the colleague is not required. Alternatively, the communication could require a response, which would mean it is two-way communication. However, the timing and quality of the response is determined by the recipient of the original correspondence.

A higher level of virtual working can be undertaken when the technologies used allow instant interaction and feedback. Technologies that allow this level of virtual working include:

- **Mobile technologies.** Most people are in constant possession of their mobile phone. This means that when a colleague or customer wishes to talk to them there is an increased chance of making direct contact. Most phones allow access to the Internet and e-mail accounts. These tools can also be accessed through laptops and netbooks. All these technologies provide the opportunity for virtual mobile working.
- **Voice over Internet Protocol (VoIP).** Whilst the original technology allowed voice communications over the Internet, more recent developments allow both voice and images to be transmitted with the use of a webcam. The ability to see and hear someone enhances the level of interaction and communication. VoIP providers such as Skype offer free use if communicating with another Skype user. This technology makes video conferences, with say colleagues around the world, an accessible and affordable activity.

- **Web 2.0 technology.** This is a collective term which covers a number of technologies. Networking sites such as MySpace and Facebook allow people with common interests to create virtual communities. These technologies, which developed out of the need for social networking, are finding their way into corporate life. Included under the Web 2.0 tag are instant messaging, blogs, podcasts and wikis. Wikis are a good example of a virtual collaborative tool as they give everyone the opportunity to create and edit information that is contained on the wiki website.

The tools and technologies of virtual working mean that there is less of a reliance on the traditional physical workspace. As long as employees have access to mobile phones, laptops or netbooks they are able to work either individually or collaboratively.

The CREAM manager needs to monitor how Web 2.0 technologies are integrated into their organizational processes, as these technologies can have an impact on the amount, type and quality of workspace required.

Embedding Performance and Productivity

The drive for greater efficiency of property provision, and ultimately cost reduction, is further fuelled by a Royal Institution of Chartered Surveyors (RICS) report, *Property in business – a waste of space?* (Bootle & Kalyan, 2002). The report claims that UK businesses are throwing away £18 billion a year through the inefficient use of space. The report proposes that whilst property is often the second highest cost after wages, it is rarely on the boardroom agenda. Whilst Bootle and Kalyan (2002) establish that £6.5 billion a year can be saved by adopting new working practices such as 'hot-desking', the main push towards new work methods appears to be based on reduced costs, rather than new work methods to improve business performance.

If the real estate or the FM operations are to be seen by the organization as more than cost-cutting departments, then it is important to demonstrate performance metrics in more than cost-cutting terms (Haynes, 2007a). Ideally the real estate and FM departments should link their performance measurements to those of the organization, thereby demonstrating the impact of the real estate and facilities on the performance of the organization.

When discussing the office environment, the terms 'physical environment' and 'behavioural environment' will be adopted. The physical environment consists of components that relate to the office occupiers' ability physically to connect with their office environment. The behavioural environment consists of components that relate to how well the office occupiers connect with each other, and the impact the office environment can have on the behaviour of individuals.

Office Productivity

The main body of literature that attempts to link office environments and productivity largely addresses the physical environment. Whilst there appears to be no universally accepted means of measuring office productivity, there does appear to be acceptance that a self-assessed measure of productivity is better then no measure of productivity (Haynes, 2007c; Leaman & Bordass, 2000; Oseland, 1999, 2004; Whitley, Dickson & Makin, 1996).

The attempts made to link the physical environment with the productivity of its occupants falls into two main categories: those of office layout (Haynes, 2008b) and office comfort (Haynes, 2008d). The literature relating to the office layout appears to revolve around two main debates: those of open plan versus cellular offices, and the matching of the office environment to the work processes.

It could be argued that the open plan debate has led to cost reduction as the prevailing paradigm with regards to office environments (Haynes, 2007d). Also, matching office environments to work processes requires a greater understanding of what people actually do when in the office environment, which is still a subject of much debate. It must be noted that much of the physical environment literature lacks any theoretical framework, and where empirical evidence was provided the sample sizes tended to be relatively small, Leaman & Bordass (2000) and Oseland (2004) being notable exceptions.

International architectural firm Gensler highlight the financial impact of poorly designed offices, claiming that they 'could be costing British business up to £135 billion every year' (Gensler, 2005, p. 1).

Gensler (2005) identified six themes from their research. A summary of these and some of the major findings are highlighted in Table 6.2.

The research by Gensler (2005) identifies the impact the office working environment has on improving productivity (potentially a 19% increase) and job satisfaction (79% of respondents linked their environment to their job satisfaction). Gensler establish a linkage between the working environment, human resources and business strategy:

> *Working environment has a fundamental impact on recruitment, retention, productivity and ultimately on the organization's ability to achieve its business strategy.*
>
> **Gensler, 2005, p. 1**

The research by Gensler (2005) was based on a survey of 200 middle and senior managers in the legal, media and financial sectors. It is acknowledged, however, that this is not a large sample size, and the sample measures the perceptions of professionals and not direct measurements of productivity (Gensler, 2005). Finally, the £135 billion cost to British businesses was based on a 19% increase in the UK service

TABLE 6.2 Summary of Gensler (2005) Research Findings

Theme	Findings
The productivity leap	A better working environment would increase employee productivity by 19%
Workplace matters	Four out of five (79%) professionals say the quality of their working environment is very important to their sense of job satisfaction
Brand control	Professionals are split 50/50 as to whether their workplace enhances their company's brand
Work styles/workspaces	Personal space (39%), climate control (24%) and daylight (21%) are the most important factors in a good working environment according to professionals surveyed
The creative office	38% of professionals believe it is difficult to be creative and innovative in their office
The 'Thinking Time' directive	78% of professionals say increasing work pressure means they have less time to think than five years ago

sector Gross Added Value. Whilst the actual value of productivity loss can be questioned, Gensler identified a clear need for research that investigates the link between productivity and office layout.

Research that attempts to address the behavioural environment tends to be at the theoretical and anecdotal stage, with little supporting empirical evidence, notable exceptions being Olson (2002) and Haynes (2005). However, there appears to be a growing awareness of the impact of the behavioural environment on office occupier productivity (Olson, 2002; Haynes, 2005). Established in the literature is the potential tension that can exist in the office environment between individual work and group work. If the office environment is to act as a conduit for knowledge creation and knowledge transfer, then offices need to allow both collaborative work and individual work to coexist without causing conflict between the two.

A wide range of papers, relating to office productivity, are collected together in *Creating the productive workplace* (Clements-Croome, 2005). The book is a valuable contribution to the office productivity debate, as it pulls together the work of a number of leading authors into a definitive key text. The book summarizes the current state of office productivity research with the chapters relating to the physical environment being more evidence based, and the chapters relating to the behavioural environment being more conceptual and anecdotal.

Behavioural Environment and Office Productivity

In this section we aim to demonstrate the role the behavioural environment plays in office productivity (Haynes, 2007d). The term *'office occupier'* will be used and is defined as a person who undertakes their work activities in an office environment. This approach enables a greater appreciation of the social context of offices. The balance between the positive interactions in the office and negative distractions will be explored. It will be established that by adopting the *'occupier perspective'* potential tensions can be identified between individual, private and team based collaborative work areas. These tensions can have an impact on the office occupier's productivity. Therefore it will be proposed that a *'people-centred'* approach to office evaluation is most appropriate for office workers with varying job tasks and allows the end user or occupier perspective to be established.

Office environments need to be designed to enhance collaboration, whilst at the same time ensuring individual private work is not compromised. Fundamentally, this section aims to explore how the office environment can affect the office occupiers' behaviour and the social environment created by office colleagues.

Behaviour and the Physical Environment

A conceptual framework, termed Servicescape, presented by Bitner (1992), aimed to establish the impact of the physical environment on the behaviour of both the customer and the employees of service organizations (Bitner, 1992). She proposes that service organizations are overlooking a valuable resource, that of the physical setting of the organization.

According to Bitner (1992) the physical environment plays such a key role in influencing buyer behaviour for service organizations that it should be integrated into the organization's marketing solution. She discusses the issue of social interactions and concludes that the physical container, the environment, affects the quality, and duration, of interactions. Whilst Bitner (1992) believes that the physical setting can affect the behaviour of its occupants, she also acknowledges that creating an environment for a range of different behaviours is a complex issue.

An important behavioural pattern acknowledged in the Servicescape framework is that of the social interaction between, and among, customers and employees (Bitner, 1992). Whilst these proposals relate to the impact of the retail environment on consumer buying behaviour, there are clearly parallels with the impact office environments can have on office occupier behaviour.

Traditionally, office evaluations have been preoccupied with the physical environment, at the expense of understanding the social environment (Stallworth & Ward, 1996). Stallworth & Ward acknowledge that research that attempts to link productivity, work setting and behaviour is a complex issue, and is in its infancy, arguing that with more and more people working in office accommodation there is clearly a business need to address the issue.

Whilst authors such as Becker (1990) and Becker & Steele (1995) argue for a non-territorial office with restricted allocation of dedicated desks, other authors aim to establish the affects of such a strategy on the office occupants (Wells, 2000). The adoption of flexible work patterns, such as 'hotelling' or 'hot desking' effectively means that employees work in a range of temporary workplaces with no particular area they can call their own (Becker, 1990). This view could overlook a behavioural need of some individuals, such as the need to express their identity and personality through the modification of their workplace environment.

Wells (2000) proposes that personalization of the work environment is a form of territorial behaviour, effectively a behaviour pattern that would be suppressed in a non-territorial office (Becker & Steele, 1995). Marquardt, Veitch & Charles (2002, p. 12) define personalization as follows: 'the process whereby workers publicly display personally meaningful items'.

Wells (2000) concludes that organizations that adopt a more lenient personalization policy report higher levels of organizational well-being. To ensure office environments work, from both an organizational and individual perspective, consideration needs to be given to the types of behaviour the office needs to enable. Increasingly, offices are becoming environments that need to create and transfer knowledge to other team members. It is this acknowledgement that has led Ward & Holtham (2000) to conclude that physical space is the most neglected resource in contemporary knowledge management.

Concentration vs. *Communication*

The possible behavioural tensions within office environments were investigated by Brenner & Cornell (1994). They specifically evaluated office environments that had been designed to enhance privacy and collaboration. The environments evaluated consisted of a small enclosed area called a *'personal harbour workspace'*, and a group area called *'common space'*. The personal harbour gave its occupants the opportunity to withdraw physically and obtain territorial privacy. The commons area consisted of group space, which was configured according to work process and technology needs (Brenner & Cornell, 2004). The environments created conformed to the commons and caves metaphor (Hurst, 1995; Steele, 1983). When people are in the 'common' areas they are available to interact with other group members, and when they wish to be on their own they can withdraw to the 'caves', thereby signalling they want privacy (Haynes, 2007d).

Brenner & Cornell (1994) investigated the willingness of the team members to trade off the need for privacy with the need for collaboration with other team members. They reported that the need for privacy diminished over the time of the experiment, and concluded that this was as a consequence of the team becoming more cohesive. Also, whilst the door on the personal harbours was not used as often as expected, it was deemed to be important by the office occupiers, as it provided them with an element of control over their environment, an issue also identified by

Leaman & Bordass (2000). The door was used to restrict their level of interaction with the other team members, and ultimately to regulate their level of privacy (Marquardt et al., 2002).

Becker & Steele (1995) reiterate the benefits of their organizational ecology concept, by claiming that it can transform physical workplace environments to support the organization's business processes. It could be argued that the connection between the workplace and the organization is central to the development of strategic facilities management. Becker & Steele (1995) propose that to ensure that the work environment supports the organization's objectives consideration needs to be given to the work processes undertaken, and the culture the organization wants to portray with its physical workspace.

Strategic FM can be developed further when emphasis is placed on the role the physical environment can play in representing organizational culture. The ultimate proposal, presented by Becker & Steele (1995), is that the physical environment can be used strategically to facilitate a change in organizational culture:

> *The planning and design of the workplace can, however, be used – or serve – as a deliberate catalyst for organizational change including the culture of the organization.*
>
> **Becker & Steele, 1995, p. 58**

Ultimately, Becker & Steele (1995) present an argument for using space to change organizational culture, which means that the physical environment will influence, and change, the patterns of behaviour in the physical environment.

Interaction vs. *Distraction*

In an attempt to create design criteria to allow the coexistence of both individual and team work, Olson (2002) created, and evaluated, a database of individual projects from multiple US-based clients between 1994 and 2000. The database contained 13,000 responses, which had been gathered by questionnaire. Olson (2002) attempts to establish the workplace qualities that have the most affect on the occupier's individual performance, team performance and job satisfaction. The workplace quality that has the strongest effect on its occupants is the ability to do distraction-free solo work (Olson, 2002). The second workplace quality to affect occupiers is support for impromptu interactions. Clearly, the tension in office environments between privacy and collaboration is brought to the surface (Haynes, 2005).

Olson (2002) argues that other people's conversations are the main interference with distraction-free working. Also, office occupiers that are in open plan or shared offices are more frequently distracted by other people's conversations than people who work in private offices. He suggests that on average office occupants spend 25% of their time making noise, such as having conversations near other people's individual workspaces. Therefore, one individual can simultaneously affect eight other office workers in a high density, open plan environment (Olson, 2002).

Whilst acknowledging the disadvantages of people having conversations in the workplace, Olson (2002) also establishes the advantages of impromptu interactions. Results from the total dataset show that 87% of respondents believe that they learn through informal interaction such as casual conversations and impromptu problem-solving sessions. However, he suggests that the scores for informal learning from both private offices and open plan offices are very similar, and therefore concludes that the idea that people in open plan environments can learn more by overhearing other people's conversations may need to be questioned.

In contrast to Olson's (2002) findings, Sims (2000) presents findings from a case study evaluation that deliberately designed space around teams, with the intention of increasing team communication and shared learning. It was called *'creative eaves-dropping'*, and it was claimed that by adopting such a team-centred approach, cycle times were reduced by 25%. In addition, the space required for the teams reduced by 43% (Sims, 2000). A limitation of this research is that whilst headline figures are presented, the research data are not provided.

Olson (2002) proposes:

> *Quiet, individual work and frequent, informal interactions are the two most time-consuming workplace activities and are the two with the greatest effects on performance and satisfaction.*
>
> Olson, 2002, p. 46

Finally, Olson (2002) suggests the answer to the potential tension between interaction and distraction is to create office environments that offer a high degree of enclosure, such as private offices. Whilst this proposal may address one side of the equation, the issue of distraction-free work, Olson (2002) does not appear to offer a solution for the other side of the equation, that environments should allow informal interaction.

The issue of distraction in the workplace is specifically addressed by Mawson (2002). He argues that anything that takes attention away from the task in hand is effectively a distraction, and therefore impacts on the performance of the individual. Mawson (2002) develops the argument by suggesting that when individuals are focused on an individual task they are in a flow state, and when they are distracted they are brought out of that flow state. The concept of workflow can be traced to DeMarco & Lister (1987). They propose that there is a time requirement for an individual to reach a deep level of concentration, termed *'ramp-up'*. If the individual is distracted, then their flow of concentration will be broken, therefore requiring further ramp-up time to reach the same level of concentration previously attained (DeMarco & Lister, 1987). Mawson (2002) argues that over the period of a day, the cumulative effect of all the distractions leads to a disruptive and less productive day. Cornell (2003) also supports the concept of workflow, highlighting that:

> *The optimal experience of flow is achieved when nearly all resources are concentrated on one task.*

Cornell (2003) proposes that to achieve optimal flow state, distractions need to be kept to a minimum. The concept of workflow, as presented by Mawson (2002) and Cornell (2003), appears to suggest that productive work is only achieved when individuals work alone. The main conclusion drawn is that the office environment needs to be a distraction-free work environment. This stance does not acknowledge different personality types, and assumes one work process, i.e. individual. The major limitation of this conclusion is that it does not acknowledge the benefits that can be obtained from different work processes, i.e. team and collaborative work (Haynes, 2005).

Factor analysis was the main statistical technique used by Haynes (2007b) to develop an understanding of the underlying concepts of office productivity. The application of factor analysis allowed 27 evaluative variables to be reduced to four distinct components (Haynes, 2007b). The results of the analysis can be seen in Table 6.3.

The components Comfort and Office Layout represent the physical environment, and the components Interaction and Distraction represent the behavioural environment (Haynes, 2007b).

TABLE 6.3 Four Components of Office Productivity and Associated Reliability

Factor	Name	Attributes	Cronbach's alpha
All			0.95
1	Comfort	Ventilation, heating, natural lighting, artificial lighting, décor, cleanliness, overall comfort, physical security	0.89
2	Office layout	Informal meeting areas, formal meeting areas, quiet areas, privacy, personal storage, general storage, work area – desk and circulation space	0.89
3	Interaction	Social interaction, work interaction, creative physical environment, overall atmosphere, position relative to colleagues, position relative to equipment, overall office layout and refreshment	0.88
4	Distraction	Interruptions, crowding, noise	0.80

The four components of office productivity were subsequently used to form a scale against which the results could be measured. The results shown in Table 6.4 identify the total positive results for Office Layout to be 28% (6 + 22) and the total positive results for Comfort to be 28% (5 + 23). It appears that the basic requirements of layout and comfort are not being addressed, which means that opportunities for productivity improvement exist by addressing the physical environment. These findings generally support the office productivity literature that has linked the physical environment to office occupier productivity (Leaman & Bordass, 2000; Oseland, 1999, 2004; Whitley et al., 1996).

The behavioural components Interaction and Distraction appear to be having the most effect on perceived productivity. The results indicate that it is the interaction component that is perceived to be having the most total positive effect (40%) on productivity, which supports the proposition that office environments are partly knowledge exchange centres (Becker & Steele, 1995). This result demonstrates that office occupiers value interaction at both a work and social level (Heerwagen, Kampschroer, Powell, & Loftness, 2004). The behavioural component Distraction has the most total negative effect (53%) on perceived productivity (Mawson, 2002; Olson, 2002). In contrast to Olson (2002) and Mawson (2002), research undertaken by Haynes (2007b) measures distraction using a multi-item scale, thereby providing a richer understanding to the distraction concept.

Clearly the distraction and interaction components are related: one person's interaction is another person's distraction. The interaction and distraction components contribute to the debate because they establish an understanding of the behavioural environment within an office environment. The challenge for managers of office environments is to maximize the interaction component, whilst at the same time attempting to minimize the distraction component. The solution to this paradox will be a combination of office work processes, office layouts, office protocols and organizational culture (Haynes, 2008a; Peterson & Beard, 2004).

TABLE 6.4 Results for Four Factors of Office Productivity (percentage response)

	Comfort	Office layout	Distraction	Interaction
Very positive	5	6	4	7
Positive	23	22	11	33
Neutral	33	34	32	35
Negative	21	22	32	16
Very negative	16	14	21	8

Open Plan vs. *Cellular Offices*

BOSTI associates, led by Michael Brill, have undertaken two major pieces of research into the effects of the workplace on worker performance. The first piece of research took place in the 1980s and collected data from 10,000 workers in 100 organizations. The findings of this study were published in a two-volume publication entitled *Using office design to increase productivity* (Brill, Marguilis, Konar & BOSTI, 1985). The second piece of research took place between 1994 and 2000 and created a database of 13,000 cases (Brill, Weidemann & BOSTI, 2001). This second wave of research acknowledged that much had changed. The four main trends that drive workplace changes were identified as (Brill et al., 2001, p. 5):

- organizational structure and strategies;
- workforce attitudes and expectations;
- technology – its ever-increasing power and widespread deployment;
- new recognitions about, and strategies for, the workplace.

Included in the second piece of research were evaluations of the individual performance, team performance and job satisfaction. With regards to office setting, the study collected data on single-occupant rooms, double-occupant rooms and open plan offices. In addition, Brill et al. (2001, p. 17) proposed some useful definitions for their research:

> *Workplace: A general term for the entire physical environment for work … the whole floor, whole building, and whole campus. The work-place always contains large numbers of workspaces.*
> *Workspace: The space where an employee sits (mostly) when in the office.*
> *Private (Cellular) Office: A workspace that has four walls to the ceiling and a door.*
> *Open (Plan) Office: A workspace whose perimeter boundaries do not go to the ceiling.*

Brill et al. (2001, p. 19) proposed that analysis of the full dataset identified ten of the most important workplace qualities in rank order. The ten workplace qualities can be seen in Table 6.5.

The top two workplace qualities relate to the specific work processes. Office workers want to be able to undertake distraction-free solo work but also value the opportunity to have an informal interaction with colleagues. Haynes (2007c) provided supporting evidence by identifying Distraction as the component having the most negative impact on perceived productivity and Interaction as having the most positive. Clearly there can be tensions in an office environment to allow individual private working to co-exist with collaborative team based working.

TABLE 6.5 Workplace Qualities in Rank Order

1	Ability to do distraction-free solo work
2	Support for impromptu interactions
3	Support for meetings and undistracted group work
4	Workspace comfort, ergonomics and enough space for work tools
5	Workspace side-by-side work and 'dropping in to chat'
6	Located near or can easily find co-workers
7	Workplace has good places for breaks
8	Access to needed technology
9	Quality lighting and access to daylight
10	Temperature control and air quality

Brill et al. (2001, p. 26) explored the issue of distraction further by investigating the amount of distraction by office type. Table 6.6 illustrates that increasing the number of occupants in an office environment increases the amount of reported distraction caused by other people's conversations. Becker (2004) shared the same concerns as Brill et al. (2001) with regards to open plan environments, especially open plan environments that contain cubicles:

> *Research by Michael Brill and his associates as well as our own studies show that despite all the furniture, technical and social fixes that have been tried to render cubicles more acceptable to employees, on the whole cubicles flunk.*
>
> **Becker, 2004, p. 25**

BOSTI Associates made the following claim, having analysed all the data from their vast database.

TABLE 6.6 Type of Office and Distraction by Other People's Conversations

	Rarely distracted	Frequently distracted
Single-room occupant	48%	29%
Double-room occupant	30%	52%
Open plan office	19%	65%

Adapted from Brill et al., 2001.

The really groovy, wide-open office, with folks shown interacting informally all day, is a visually seductive myth. Research shows it doesn't support work very well and, in fact, can incur significant losses in individual and team performance and job satisfaction.

Brill, Keable & Fabiniak, 2000, p. 36

Balancing Collaborative and Individual Work Environments

An extensive literature review of research that attempts to establish links between the ways that knowledge workers collaborate and the physical environment provided, was undertaken by Heerwagen et al. (2004). The basis of the review was the research question: 'How can the physical design of the workplace enhance collaborations without compromising an individual's productivity?' (Heerwagen et al., 2004, p. 510).

Heerwagen et al. (2004) define the nature of knowledge workers as being a combination of high cognitive skills and social interaction. They develop the argument to suggest that there are two basic needs of knowledge workers. They are:

- time to work alone to think, analyse and reflect; and
- time to interact with others so that ideas can be generated and evaluated.

In common with Mawson (2002) and Cornell (2003), Heerwagen et al. (2004) acknowledge the benefits of private individual work, although, in contrast, they also report the benefits of, and the need for, collaborative work of knowledge workers.

It is proposed that collaborative knowledge work consists of two dimensions: the social dimension and the individual dimension (Heerwagen et al., 2004). They also propose that the social dimension can be subdivided into three components, with each component being dependent on the amount of time spent with colleagues. The three components are:

- awareness;
- brief interaction;
- collaboration.

The awareness component relates to the eavesdropping concept presented by Sims (2000), the idea that office occupiers have a general awareness of what is going on in the office environment just by overhearing office conversations. Heerwagen et al. (2004) propose that the key physical requirements to ensure that the awareness dimension is supported are visual and aural accessibility. The physical requirement proposed is that of a highly open environment. Heerwagen et al. (2004) acknowledge the potential problems of a high-awareness environment as being loss of privacy, loss of confidentiality, distraction and interruptions, although they argue that in an open environment interruptions and distractions may be reduced because of non-verbal and behavioural cues.

When people are focused on an individual task, their posture, eye gaze and demeanour indicate they are not available for conversation. However, if they look up, make eye contact or walk around, others are more likely to perceive them as available for interaction.

(Heerwagen et al., 2004, p. 514)

Whilst some people may observe the behavioural cues for interaction others may not, therefore to ensure interruptions and distractions are kept to a minimum office protocols would need to be introduced (Brennan, Chugh & Kline, 2002; Sims, 2000).

The benefits to the knowledge worker of ad hoc brief interactions with colleagues are identified by Heerwagen et al. (2004). According to Heerwagen et al. (2004) brief interactions can be both intentional and unintentional, and can occur in many locations, i.e. at people's desks, in the corridor and near central services. The location of the brief interaction can be considered an 'information exchange' (Heerwagen et al., 2004). They present the argument that the important predictors of interaction are layout and circulation. The visibility or *'line of sight'* within an office environment can influence the amount of interaction within the office (Heerwagen et al., 2004).

The patterns of interaction between colleagues are greater for open environments than closed office environments (Becker & Sims, 2001). Therefore open plan environments can enhance social behaviour by enabling increased opportunities for communication, and in the building of relationships with colleagues (Becker & Sims, 2001). Haynes (2007b) proposes that the creation of *'informal interaction points'* such as tea points, printing and copying facilities can facilitate chance conversations, and thereby enable knowledge creation and knowledge transfer.

Reviewing the literature that aims to establish links between individual work and physical space, Heerwagen et al. (2004) establish the benefits of individual workspaces that support focused concentration by reducing distractions and interruptions. They acknowledge that providing this type of environment is in tension with the desire to create an environment that enables interaction. In evaluating the tension between collaborative and individual work Heerwagen et al. (2004) warn that a balance needs to be achieved.

When office occupiers need to concentrate they should be able to move to quiet spaces elsewhere in the office (Becker & Sims, 2001). This assumes that the office worker has the flexibility to leave their desk and the autonomy to work flexibly (Laing, Duffy, Jaunzens & Willis, 1998).

Creating collaborative office environments requires the integration of both the social and the individual factors; however there is little research evidence that links collaborative behaviour to the physical space (Heerwagen et al., 2004). A piece of research that attempts to evaluate the impact of the workplace environment on the individual's privacy and team interaction is presented by Peterson & Beard (2004). They acknowledge that little independent research has been undertaken on new workplace designs, and that work environment manufacturers have largely funded research that has been undertaken. Therefore Peterson & Beard (2004) evaluated a new work environment, which had been designed to include commons and personal harbours, in response to the need for both interaction and privacy. The design was based on the caves and commons metaphor (Steele, 1983).

Peterson & Beard (2004) report that with regards to the individual workspace, i.e. the personal harbours, the participants reported that they were satisfied with the visual privacy that it provided, and the ability to concentrate, and the amount and quality of work they could accomplish. However, participants did report that they were not satisfied with the auditory privacy of the individual workspace. Peterson & Beard (2004) acknowledge that the results appear contradictory: participants report that they are not satisfied with the auditory privacy, such as noise levels, yet are satisfied with their ability to perform their work. Peterson & Beard (2004) explain that the doors on the personal harbours contain a white noise system, so that when the door is shut all background noise is eliminated. However, observation of the office work methods revealed that people did not close the harbour door, therefore not activating the white noise system, and consequently allowing noise from the common areas into the individual workspace. The door was provided for the office occupiers as a means of regulating their interaction (Marquardt et al., 2002), although the occupiers were not exercising that option.

The respondents reported general satisfaction with the group area, i.e. commons (Peterson & Beard, 2004). Specifically, they were satisfied with the access and interaction with other group members. They were also satisfied with the quality, and amount of work the group accomplished. One issue that the respondents reported some dissatisfaction with was the lack of a suitable area for the display of group information, an issue previously identified by Becker & Steele (1995):

> *Displayed thinking, especially using the simple anonymous feedback medium of Post-It notes, allows people to challenge ideas and suggest new ones without fear of confrontation.*
>
> **Becker & Steel, 1995, p. 82**

Peterson & Beard (2004) finally conclude that using the working environment to enable team collaboration and communication also leads to team cohesiveness. The results indicate that 85% of the participants reported feeling a sense of closeness and camaraderie with other team members, indicating the behavioural component of the office environment (Peterson & Beard, 2004).

Occupier Perspective

Understanding how people react to the built environment is at the centre of Environment-Behaviour Studies (EBS). To develop the debate there is a requirement to establish the occupier perspective. Rapoport (1990) identifies the importance of *'meaning'* to the users of the built environment, and goes as far as to differentiate the users' meaning from the designers' meaning.

> *One of the hallmarks of man–environment research is the realization that designers and users are very different in their reactions to environments, their preferences, and so on, partly because their schemata vary. It is thus users' meaning that is important not architects' or critics'; it is the meaning of every day environments, not famous buildings – historical or modern.*
>
> **Rapoport, 1990, pp. 15-16**

In an attempt to address the differing needs of office occupiers Fleming (2004) proposes a conceptual framework. He proposes that assessment of work environments should include the occupier perspective. Subsequently, he develops an argument for behavioural assessment of work environments to complement the physical assessments.

> *The mechanistic, quantitative nature of building performance paradigms fails to take into account the effect of occupiers' perceptions of their environments. Facility managers currently see buildings as containers of products and not containers of people. Products are measured against technical performance specifications rather than the idiosyncratic thoughts and perceptions of the building occupants.*
>
> **Fleming, 2004, p. 35**

This approach contributes to the debate by proposing a conceptual framework that establishes that traditional property performance has largely concentrated on the alpha press measures, such as observations by detached non-participants (Fleming, 2004). It develops the argument by proposing that a greater understanding of the behavioural environment can be obtained by the use of beta press measures; the occupier perspective.

Support for Fleming's (2004) call for a paradigm shift with regards to evaluation of office environments is found with Duffy (2000) and Haynes (2007c). Duffy (2000) proposes that office environments have changed relatively little over the past 20 years. Duffy (2000) attributes the lack of development to a preoccupation with hierarchical cultures, Taylorist mentalities and a cost reduction emphasis (Duffy, 2000). Haynes (2007b) argues that traditional office research has tended to adopt a purely rationalist paradigm, with the missing component for a theoretical framework being the consideration of the behavioural environment. A possible way of understanding the behavioural environment would be to consider the connectivity that takes place in an office environment.

In an attempt to propose a new research paradigm for office productivity evaluations, Haynes (2005) develops the concepts of *'buildings ecology'* (Levin, 1981) and *'office ecology'* (Becker, 1990), and proposes the concept of *'workplace connectivity'*. The principle is that to truly evaluate the impact of the office environment on occupiers' productivity, there needs to be an understanding of the connectivity between the office occupiers and their work environment. The occupier connectivity with their environment consists of:

- a psychological (perceptual) response
- a physiological (biological) response

It is proposed that the psychological (perceptual) response of the office occupier to their office environment is an area that requires further research, thereby enabling a better understanding of office productivity (Haynes, 2007d). In addition, further research could be based on a theoretical framework which includes the different levels of connectivity that office occupiers have with their organization.

Proposed Theoretical Framework for Connectivity

The proposed theoretical framework aims to acknowledge the psychological and physiological response that office occupiers have with their work environment (Haynes, 2007c, 2008e). The connectivity of the office occupier with their work environment can be considered at three different levels, as can be seen in Figure 6.3. The first level of connectivity is the workplace connectivity. This level relates to the

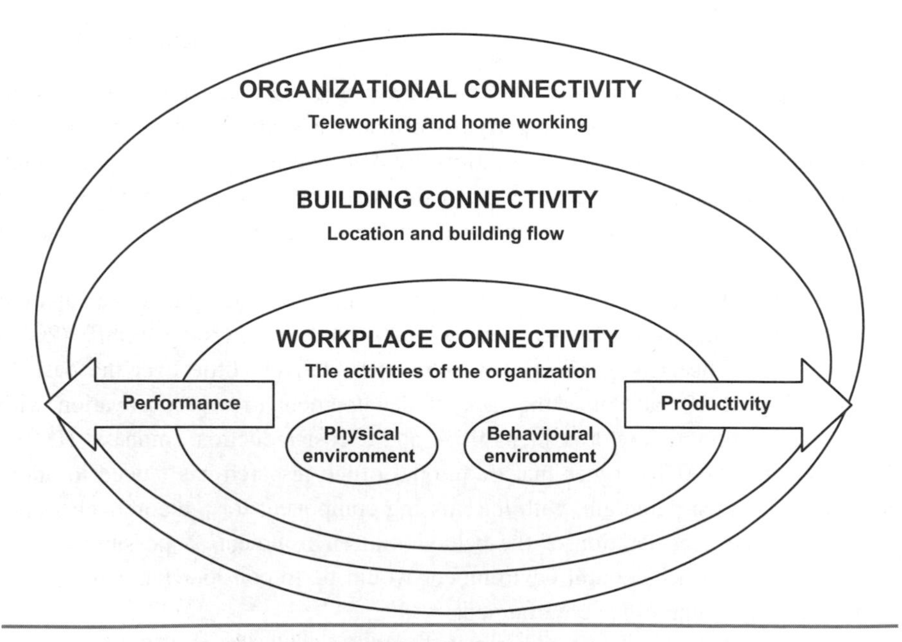

Figure 6.3 Proposed theoretical framework for levels of connectivity. *(From Haynes, 2008e; reproduced with permission)*

office environment and the impact the behavioural and physical environment have on perceived productivity. The second level is building connectivity. This relates to the location of the building, its connection with the local environment and also to the flow of occupiers within the building. The final level is the organizational connectivity. This relates to how people connect to the organization when not in the office environment. This level of connectivity could be classed as *'organizational glue'*.

Conclusion

The complexity of researching occupier productivity and office environments has been explored. The complexity comes from the proposal that the productivity of the office occupier is not only dependent on the physical environment, but also the behavioural environment. The addition of the social context has subsequently meant that the definition of office productivity has remained ill defined. We believe that the contemporary corporate real estate asset manager needs to engage with this research to capture productivity benefits from the CREAM standpoint.

It is proposed that future office productivity research needs to adopt an office occupier perspective. The adoption of this stance would reveal how office occupiers make sense of their work environment, and how they attempt to create a sense of belonging in their work environment. The occupier perspective also offers the opportunity to identify a potential tension between individual private work and team based collaborative work. A review of the office environment literature has established that this is an area that requires further research.

The future for office and workplace productivity measurement is to establish links between real estate and facilities performance metrics and the organizational performance metrics. Establishing these links will demonstrate a strategic integration between organizational demand and the provision of facilities and real estate solutions.

Summary Checklist

1. For your organization determine how the workplace has become distributed. This should include evaluations of:
   a. Work undertaken in the office and the different work processes adopted.
   b. Work undertaken out of the office and the range of locations utilized.
   c. Work undertaken virtually through the use of information and communication technologies (ICT).
2. Where open plan office environments are used, adopt strategies that aim to maximize the advantages of open plan whilst at the same time minimizing the potential disadvantages of open plan.

3. Ensure that detailed attention is given to lighting, temperature and ventilation if the impacts of the physical attributes of the office environment are to have a positive impact on office occupier productivity.
4. It is important that the CREAM manager pays specific attention to the behavioural office environment. The office spaces created should be managed to optimize the balance between interaction and distraction. This can be achieved by including a range of appropriate spaces specifically designed to support:
   a. group collaborative working; and
   b. private individual working.
5. The CREAM manager should ensure that any workplace solution enhance the three levels of connectivity. The levels are:
   a. Workplace connectivity. The connection between the office occupier and their working environment.
   b. Building connectivity. This relates to the way that people move around the building. This also includes the physical location of the building.
   c. Organizational connectivity. This includes the systems, technologies, procedures and protocols used to maintain connection with people when they are not working in the office environment.

References

Becker, F. (1990). *The total workplace: Facilities management and the elastic organization*. New York, NY: Van Nostrand Reinhold.

Becker, F. (2004). *Offices at work*. San Francisco, CA: Jossey–Bass.

Becker, F., & Sims, W. (2001). *Offices that work: Balancing communications, flexibility and cost*. Cornell University, International Workplace Studies program.

Becker, F., & Steele, F. (1995). *Workplace by design: Mapping the high-performance workscape*. San Francisco, CA: Jossey–Bass.

Bitner, M. J. (1992). Servicescapes: the impact of physical surroundings on customers and employees. *Journal of Marketing, 56* (April), 57–71.

Bootle, R., & Kalyan, S. (2002). *Property in business – a waste of space?* London: RICS.

Brennan, A., Chugh, J., & Kline, T. (2002). Traditional versus open office design: a longitudinal field study. *Environment and Behavior, 34*(3), 279–299.

Brenner, P., & Cornell, P. (1994). The balance between privacy and collaboration in knowledge worker teams. In: *Proceedings of the Fourth International Symposium on Human Factors in Organizational Design and Management*. Sweden: Stockholm. 29 May–2 June.

Brill, M., Keable, E., & Fabiniak, J. (2000). The myth of open plan. *Facilities Design & Management, 19* (2), 36–39.

Brill, M., Marguilis, S., Konar, E., & BOSTI. (1985). *Using office design to increase productivity,* Vols. 1 and 2. Buffalo, NY: Workplace Design and Productivity Inc.

Brill, M., Weidemann, S., & BOSTI. (2001). *Disproving widespread myths about workplace design*. Buffalo, NY: Kimball International.

Clements-Croome. (2005). *Creating the productive workplace* (2nd ed.). London: Routledge.

Cornell, P. (2003). *Go with the flow.* Retrieved from Steelcase. http://www.steelcase.com/Files/027c3093ec534a79bf62aa77d71a0cbb(1). pdf Accessed 15.09.09.

DeMarco, T., & Lister, T. (1987). *Peopleware: Productive projects and teams.* New York, NY: Dorset House.

Drucker, P. (1959). *Landmarks of tomorrow: A report on the new 'post-modern' world.* New York, NY: Harper.

Duffy, F. (1992). *The changing workplace.* London: Phaidon Press.

Duffy, F. (2000). Design and facilities management in a time of change. *Facilities, 18*(10/11/12), 371–375.

Fleming, D. (2004). Facilities management: a behavioural approach. *Facilities, 22*(1/2), 35–43.

Gensler. (2005). *These four walls: The real British office*. London: Gensler.

Gensler. (2008a). *2008 Workplace Survey: United Kingdom.* London: Gensler.

Gensler. (2008b). *2008 Workplace Survey: United States.* San Francisco, CA: Gensler.

Gibson, V. (2001). In search of flexibility in corporate real estate portfolios. *Journal of Corporate Real Estate, 3*(1), 38–45.

Handy, C. (1989). *The age of unreason.* London: Business Books.

Handy, C. (1994). *The empty raincoat: Making sense of the future.* London: Hutchinson.

Hardy, B., Graham, R., Stansall, P., White, A., Harrison, A., Bell, A., et al. (2008). *Working beyond walls: The government workplace as an agent of change*. London: DEGW and OGC.

Haynes, B. P. (2005). *Workplace connectivity: a study of its impact on self-assessed productivity.* PhD Thesis. UK: Sheffield Hallam University.

Haynes, B. P. (2007a). Office productivity: a shift from cost reduction to human contribution. *Facilities, 25*(11/12), 452–462.

Haynes, B. P. (2007b). Office productivity: a theoretical framework. *Journal of Corporate Real Estate, 9* (2), 97–110.

Haynes, B. P. (2007c). An evaluation of office productivity measurement. *Journal of Corporate Real Estate, 9*(3), 144–155.

Haynes, B. P. (2007d). The impact of the behavioural environment on office productivity. *Journal of Facilities Management, 5*(3), 158–171.

Haynes, B. P. (2008a). Impact of workplace connectivity on office productivity. *Journal of Corporate Real Estate, 10*(4), 286–302.

Haynes, B. P. (2008b). The impact of office layout on productivity. *Journal of Facilities Management, 6* (3), 189–201.

Haynes, B. P. (2008c). An evaluation of the impact of office environment on productivity. *Facilities, 26* (5/6), 178–195.

Haynes, B. P. (2008d). The impact of office comfort on productivity. *Journal of Facilities Management, 6* (1), 35–51.

Haynes, B. P. (2008e). The role of the behavioural environment in enhancing office productivity. In E. Finch, & D. Then (Eds.), *CIB W70 Conference Proceedings* (pp. 451–461). Edinburgh, Scotland: Heriot–Watt University, 15–18 June, CIB W70.

Heerwagen, J. H., Kampschroer, K., Powell, K. M., & Loftness, V. (2004). Collaborative knowledge work environments. *Building Research and Information, 32*(6), 510–528.

Hurst, D. K. (1995). *Crisis and renewal: Meeting the challenge of organizational change*. Cambridge, MA: Harvard Business Press.

Jones Lang LaSalle. (2008). *Perspectives on workplace*. London: Jones Lang LaSalle.

Laing, A., Duffy, F., Jaunzens, D., & Willis, S. (1998). *New environments for working: The redesign of offices and the environmental systems for new ways of working*. London: E & FN Spon.

Leaman, A., & Bordass, W. (2000). Productivity in buildings: the 'killer' variables. In D. Clements-Croome (Ed.), *Creating the productive workplace* (pp. 167–191). London: E & FN Spon.

Levin, H. (1981). Building ecology. *Progressive Architecture, 62*(4), 173–175.

Marquardt, C. J., Veitch, J. A., & Charles, K. E. (2002). *Environmental satisfaction with open-plan office furniture design and layout.* Institute for Research in Construction. Ottawa: National Research Council of Canada. RR–106.

Mawson, A. (2002). *The workplace and its impact on productivity.* Retrieved from Advanced Workplace Associates. http://www.advanced-workplace.com/advanced_working_papers/research.php?id=34. Accessed 18.09.09.

Olson, J. (2002). Research about office workplace activities important to US business – and how to support them. *Journal of Facilities Management, 1*(1), 31–47.

Oseland, N. (1999). *Environmental factors affecting office worker performance: A review of evidence.* Technical Memoranda TM24. London: CIBE.

Oseland, N. (2004). *Occupant feedback tools of the office productivity network.* Retrieved June 2004 from Office Productivity Network. http://www.officeproductivity.co.uk.

Peterson, T. O., & Beard, J. W. (2004). Workplace technology's impact on individual privacy and team interaction. *Team Performance Management, 10*(7/8), 163–172.

Rapoport, A. (1990). *Meaning of the built environment: A non-verbal communication approach.* Tucson, AZ: University of Arizona Press.

Roethlisberger, F. J., & Dickson, W. J. (1939). *Management and the worker.* Cambridge, MA: Harvard University Press.

Sims, W. (2000). Team space: planning and managing environments to support team work. *International Journal of Facilities Management, 1*(1), 21–33.

Smith, J. (2008). *Welcoming Workplace.* London: Helen Hamlyn Centre, Royal College of Art.

Stallworth, O. E., & Ward, L. M. (1996). Recent developments in office design. *Facilities, 14*(1/2), 34–42.

Steele, F. (1983). The ecology of executive teams: a new view of the top. *Organizational Dynamics* (Spring), 65–78.

Ward, V., & Holtham, C. (2000). *The role of private and public spaces in knowledge management.* Retrieved from Knowledge Management: Concepts and Controversies, Conference proceedings, University of Warwick, Coventry. http://www.sparknow.net/page_attachments/0000/0082/Public_Spaces_in_KM.pdf. Accessed 18.09.09.

Wells, M. M. (2000). Office clutter of meaningful personal displays: the role of office personalization in employee and organizational well-being. *Journal of Environmental Psychology, 20*, 239–255.

Whitley, T. D., Dickson, D. J., & Makin, P. J. (1996). The contribution of occupational and organizational psychology to the understanding of sick building syndrome. In: *Proceedings CIBSE/ASHRAE Joint National Conference* (pp. 133–138), 28 September–1 October, Harrogate, UK.

Chapter 7

Corporate Relocation

Introduction

In the previous chapter we established the changing purpose of the office environment with the shift in emphasis from the traditional office, with individual workers undertaking process work, to a more collaborative dynamic environment with people interacting on an ad hoc basis. The space created in the office environment can have a direct impact on the creativity, productivity and the health and wellbeing of its occupants. Using the workspace to deliver an organization's business objectives raises the space planning process to a strategic level and is a good example of how the CREAM strategic planning process can link directly to the corporate strategic planning process.

The tensions and pressures facing many organizations mean that they have to respond to uncertainties such as business drivers for change, competitors and new entrants to the market (see Chapter 1). Creating an accommodation strategy for a constantly changing business environment requires a methodology that can anticipate such uncertainty.

In this chapter we will outline the strategic space planning alignment process which aims to ensure the accommodation and workplace provision meets the demands of the organization.

Strategic Space Planning Alignment Process

Essentially, strategic space planning can be described in simple terms as:

> *The identification and translation of the organizational and business need for accommodation space into cost effective space solutions.*

The starting point for the strategic space planning process is the identification of the organizational demand. Adopting this approach requires an understanding of the occupier requirements and an evaluation of the usability of any proposed buildings.

Corporate Real Estate Asset Management. ISBN: 978-0-08-096680-9

This means that buildings are ultimately designed from the inside out with the outside appearance and the external architecture of the buildings acting as an envelope to the organization. To ensure that buildings are designed for their occupants and usability, it is essential that user representatives are involved at the earliest stages of space planning. Schramm (2005, p. 31) makes the observation:

> *In general, decision-making about buildings and space poses the risk of questioning, if not rejection, on the part of the organization's employees, if it is imposed as a top-down process.*

It is important to identify that space planning can only truly become strategic when it is linked to the corporate objectives and hence the business plan of the organization. Organizational space is a resource like any other organizational resource; it requires that same amount of attention with regards to planning and monitoring. It should therefore be fully integrated with the corporate planning process.

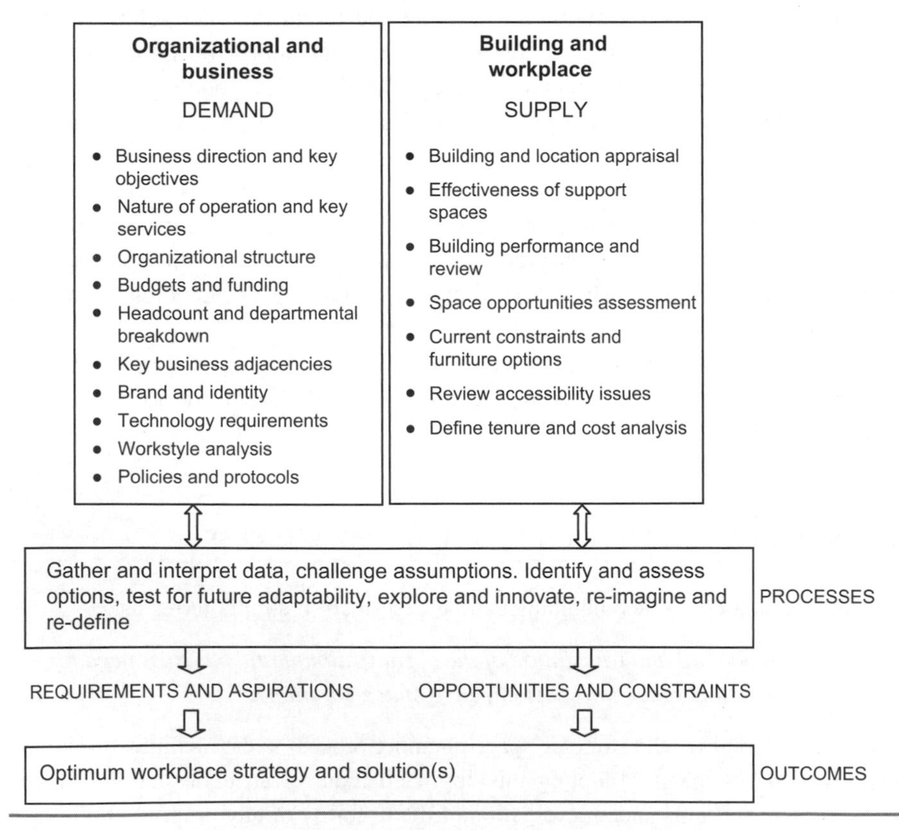

Figure 7.1 Demand and supply diagram. *(From Hardy et al., 2008, p. 85. © Crown Copyright)*

Strategic space planning can be useful if an organization wants to undertake a major change. Such business changes could include expansion or contraction of staff numbers. These types of changes would lead to an evaluation of the appropriateness of the current accommodation provision. This may lead to decisions about refurbishment or relocation.

The basic methodology of matching organizational space demand with the possible accommodation and workspace solutions provides organizations with a meaningful way of evaluating their buildings to determine their relative capacity to accommodate and support the organization and its business plan. A useful model that clearly illustrates the relationship between supply and demand is incorporated in the publication entitled *Working beyond walls – the government workplace as an agent of change*, by DEGW and the Office of Government Commerce (OGC) (Hardy et al., 2008). The model, shown in Figure 7.1, illustrates the inputs that are required from the organization and the building supply, the processes that are undertaken and the outputs that are achieved from the matching process (Hardy ct al., 2008). Adopting the organizational demand and building supply model enables a structured approach to the decision making process. This simple methodology means that the organization's demand for space is established before any buildings are considered. This includes buildings that may already be in the organization's property portfolio.

Establishing Organizational and Business Demand

The first stage of the strategic space planning alignment process is the identification of the amount of space required to support the organization. A methodology is adopted that starts with the macro and business requirements for space and includes a number of stages that lead to the evaluation and analysis of work styles undertaken within the organization. The following section will explore the components incorporated in establishing organizational and business demand (Hardy et al., 2008), and will expand upon the evaluations and analytical techniques required for each stage of the process.

Business Direction and Key Objectives

Ensuring that the accommodation and workplace solutions completely align to the organizational demand requires the CREAM manager to completely understand the aims and direction of the business. This is a clear example of where the CREAM manager has to be able to crossover discipline boundaries. The CREAM manager has to have the appropriate business skills to be effective in supporting the aims of the business.

Before an organization can establish its direction and aims it would need to undertake a strategic analysis of its business environment. The evaluative techniques that can be used have been discussed in Chapter 1, but could include analytical tools such as:

- political, economic, social, environmental and legal (PESTEL) analysis;
- mission and vision statements;
- strengths, weaknesses, opportunities and threats (SWOT) analysis or Scored SWOT analysis;
- competitor analysis;
- scenario planning.

It is tools and techniques such as these that the CREAM manager has to be able to understand, and interpret, if they are to perform at a strategic level within the organization.

Once the context that business decisions are made in has been established, it is important to understand the challenges facing an organization and the corporate strategy it wishes to adopt. In Chapter 2 we discussed the possible ways that the CREAM strategy could align with the corporate strategy. Achieving this alignment in the context of accommodation provision means that the CREAM manager has to collect enough data and information to inform the decision making process. The data collection and analytical techniques are summarized below:

- Document analysis. This would entail a detailed evaluation of the corporate strategy contained in the organization's business plan. An organization's vision or mission statement should establish the direction of the business.
- Interviews. This would include interviews with the CEO of the organization and all the other functional directors. The individual interviews would help establish how the different functions of the organization intend to support the corporate strategy.
- Focus group. Included in the focus group would be all the key personnel that had contributed to the corporate strategy. This kind of information enables the CREAM manager to establish the consensus view with regards to the direction of the business.

Nature of Operation and Key Services

Whilst the CREAM manager may be responsible for the buildings and services that support the core business it must be remembered that the core business is fundamentally the organization's reason for being. This being the case, the CREAM manager needs to understand the nature of the core business and the essential services that are part of its unique selling point. If the core business is marketing or some kind of media production then it would be important for the CREAM manager to understand how marketing campaigns are compiled or how media are created.

Establishing a deeper understanding of the nature of the business allows the CREAM manager to develop an accommodation and workspace strategy that best supports, and even enhances, the core business processes. Types of data gathering and analytical techniques that would assist the CREAM manager in understanding the nature of the operation and key services are summarized below:

- Document analysis. An evaluation of the marketing materials for the organization. This would help identify the products and services the organization offers.
- Interviews. Undertaking interviews with key senior departmental managers would help establish more detail about how they develop the products and services that the organization offers.

Organizational Structure

An organization should aim to structure itself in the best way possible to deliver its corporate strategy. Determining the structure of an organization may include decisions about centralization or decentralization. The former may lead to larger departments with increased headcounts, whilst the latter may lead to smaller business units. The departments with a larger number of occupants may wish to have standardized space allocation, whereas the smaller business units may wish to have space that meets their specific needs.

The number of levels in the organizational structure can also send signals to the CREAM manager about the management style the organization wishes to adopt. A flat organizational structure could mean that a collaborative democratic management style is adopted. Many levels of management could signal multiple lines of authority and a command and control management style. The CREAM manager needs to establish the management styles so that they can provide workspace that reiterates and enforces that of the organization. The main data gathering and analytical technique required to evaluate the organizational structure is:

- Document analysis. An evaluation of the organizational chart will help to identify the reporting structure and levels of hierarchy in the organization.

Cultural Aspirations

As we previously discussed in Chapter 2, the accommodation and workspace provided for employees can send clear signals about the culture of the organization. Organizations may see an office relocation, or refurbishment, as an opportune time to change the organizational culture at the same time as changing the workspace provision. It is important the CREAM manager establishes an understanding and appreciation of the softer aspects of organizations. These softer aspects include the beliefs and values of an organization, which may be implicit or explicit. Identifying the organizational culture is like getting underneath the skin of the organization and establishing what it feels like to be an employee or a customer/client of the organization.

If an organization wishes to undertake a cultural change then the CREAM manager must consider how the accommodation and workspace provided can facilitate that change. If the organization has the cultural aspirations to be a more open and collaborative culture, then it would be appropriate for the CREAM manager to provide interactive work areas and space standards that support these aspirations.

The main data gathering and analytical technique required to evaluate the cultural aspirations are summarized below:

- Interviews. This would include individual interviews with the CEO of the organization and all the other functional directors. It is important to establish how the different functional leaders see the organizational culture, and to establish any nuances or contradictions.
- Focus group. Included in the focus group would be all the key strategic personnel. The focus group would establish a consensus view with regards to the cultural aspirations of the organization.
- Cultural web analysis. The cultural web analysis is an evaluation of six elements which collectively form the paradigm of the organization (Johnson & Scholes, 1999). An evaluation of the six elements enables a fuller picture of the organizational culture to be established (see Chapter 2).

Budgets and Funding

Establishing the organizational demand involves identifying the amount and type of space an organization needs to deliver its corporate strategy. However, the CREAM manager has to meet the organizational demand within certain boundaries. One important consideration would be the budgeting and funding of the accommodation and workplace strategy.

Whilst realistic budgets need to be established for relocation and refurbishment projects the CREAM manager needs to identify other added value performance metrics for the project. As identified in Chapter 2, if space provision is only measured in cost per square metre, the CREAM manager could be trapped in the cost reduction paradigm.

Headcount and Departmental Breakdown

Evaluating the total headcounts and that for each department gives the CREAM manager an indication of the scale of the relocation project. It is important to establish if the headcount corresponds to a full-time, part-time or a contractual worker.

The headcounts should include any forecasts for future growth or contraction. Issues of changing staffing levels and makeup should be clearly identified. In addition, it is important for the CREAM manager to know of any possible future outsourcing of staff or departments as this could significantly impact the accommodation and workplace strategy.

Forecast growth in employees does not necessarily mean that an organization needs more space. The increase in staff numbers could possibly be absorbed by new working methods and more mobile working.

The main data gathering and analytical technique required to evaluate the headcount and departmental breakdown are summarized below:

- Document analysis. An evaluation of the organizational chart will help to identify the departments and the number of employees in the departments. It is important to include headcount forecasts.

Key Business Adjacencies

Evaluating organizational charts can help in establishing hierarchies, reporting lines, departmental structures and staffing levels but they do not always portray how the organization actually works. It is important to establish how the departments interact with each other.

The relationships between departments/functions can be identified by the use of an adjacency diagram. Figure 7.2 shows an illustrative example of an adjacency diagram. The principles adopted in creating such a diagram are summarized below:

- Each function is represented by a sphere/oval.
- The size of the sphere/oval is proportionate to the number of employees in that function.
- The strength of the connecting line is representative of the strength of relationship between departments.
  - A thick line indicates a very strong working relationship between departments.
  - A normal line indicates an average relationship between departments.
  - A thin line indicates a weak relationship between departments.
  - Finally, a dotted line indicates an intermittent relationship between departments.

The main data gathering and analytical techniques required to evaluate the key business adjacencies are summarized below.

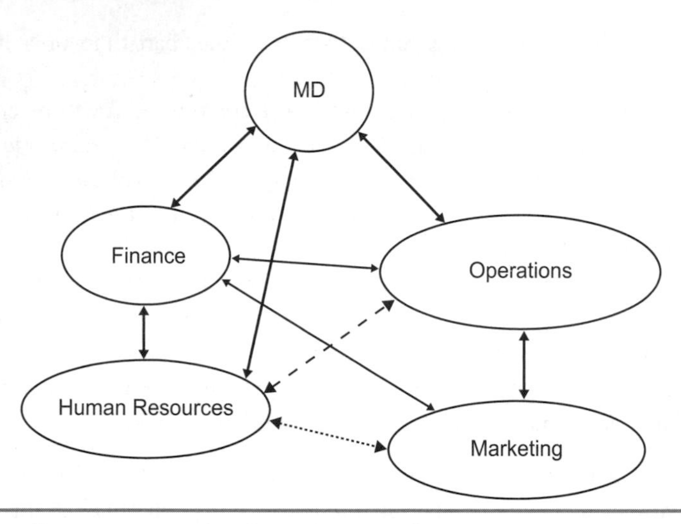

Figure 7.2 Illustrative example of an adjacency diagram

- Document analysis. An evaluation of the organizational chart will help to identify the departments and the number of employees in the departments.
- Interviews. This would include representatives for the functional parts of the organization. It is important to capture how the departments actually work with each other.
- Observation. A possible technique that could be used to monitor actual physical interactions between departments/functions.
- Adjacency diagram. An evaluative tool that graphically illustrates how the functions within an organization actually interact with each other.

Brand and Identity

We discussed in Chapter 2, the brand and identity of an organization can be clearly communicated to its employees and customers, through the accommodation and workspace provision. The location of an organization's buildings, their external and internal design and appearance all send signals about the company brand and identity.

Brand and identity, in the context of the organization accommodation, can be referred to as the *'expression'*. CABE (2005) define expression as:

> *Communicating messages both to the inhabitants of the building and to those who visit it, to influence the way they think about the organization – getting the most from the brand.*
>
> **CABE, 2005, p. 7**

Whilst the value of branding can be hard to measure in a tangible form, the benefits of using office space to articulate the organizational brand identity are set to increase (CABE, 2005).

Technology Requirements

The continuing developments in information and communication technologies (ICT) mean that people can remain connected at all times. The use of such technologies means that office workers are no longer restricted to their office environments. Even when they are in the office environment employees have the flexibility to roam around the building whilst still remaining connected using wireless technology. It is just as easy to send an e-mail or surf the Internet with a wireless laptop or a mobile phone.

The adoption of more centralized ICT service provision means that data, and even programs, could be stored centrally. The knowledge worker can access their work anywhere, and at any time, as long as they have an Internet connection (Hardy et al., 2008).

It is essential to evaluate the impacts that the developments in ICT may have on the demands for physical workspace. This will require the forecasting of future ICT developments and assessing their possible take up by office workers.

The main data gathering and analytical techniques required to undertake technology requirements are summarized below:

- Site survey. To evaluate current technologies being implemented by the office workers.
- ICT forecast. An evaluation of the emerging technologies that may have an impact on office working habits and subsequent demand for physical office space.
- Cultural readiness to change. The data gathering technique adopted could be a survey with the aim of establishing office occupiers' readiness to change their working practices as a result of emerging technologies.

Workstyle Analysis

One of the central elements in a relocation project is a thorough evaluation of office occupiers' work processes. The work processes are at the heart of an organization's activities and therefore any accommodation and work space strategy should aim to facilitate and enhance the productivity of those work processes.

It is essential that the CREAM manager undertakes a workstyle analysis to establish the different kinds of work activity undertaken by the office occupiers. Once the work activities have been identified, and categorized, consideration can be given to creating an office environment that integrates the supporting work settings.

There are a number of ways that office work activities can be categorized. The work style categorizations tend to be around whether the office worker works individually or in a team, and whether the office worker is mobile inside or outside the office environment. An additional dimension that could be included in the work style analysis is the age of the office worker.

The Welcoming Workplace project categorizes three different kinds of space requirements for the ageing knowledge worker (Smith, 2008). The three categories of space requirements are summarized below:

- **Spaces to concentrate.** This kind of space provision acknowledges that knowledge work is sometimes undertaken by the individual. This means that space is required that allows a high level of concentration to be achieved without distraction.
- **Spaces to collaborate.** The creation and communication of knowledge and ideas means there needs to be space allocated for team working activities.
- **Spaces to contemplate.** People need space to take a break from the everyday activities of the office working environment. These kinds of spaces enable people to relax and refocus their thoughts.

Whilst the Welcoming Workplace project was primarily research investigating the ageing knowledge worker, the categorizations of space requirements could be considered to be appropriate for any kind of knowledge worker.

In an attempt to identify the future office workspaces for ageing knowledge workers, the criteria of collaboration and concentration were used to create the model shown in Figure 7.3 (Erlich & Bichard, 2008). The figure illustrates that the creation of project space enables increased concentration and increased collaboration. It also proposes that collaboration that does not need high concentration levels can be undertaken in spaces such as social space and meeting space. The model proposes that more individual spaces for concentration in the office environment can be created by the use of booths (Erlich & Bichard, 2008). A practical application of booths in the office environment can be seen in the Nokia case study.

Gensler, the international architectural practice, identify four work modes for knowledge workplace activities (Gensler, 2008). The four work modes can be seen in Table 7.1.

The four work modes shown in Table 7.1 are summarized below:

- **Collaborate.** This mode of working utilizes the collective intelligence of group and team working.
- **Focus.** This workstyle acknowledges the productivity gains that can be achieved by allowing an individual to work in a distraction-free environment.
- **Learn.** Knowledge workers have to acquire new ideas and information in the same way as a student at a university might. Since people's learning styles can be different, there is a need to create a range of appropriate different learning environments.

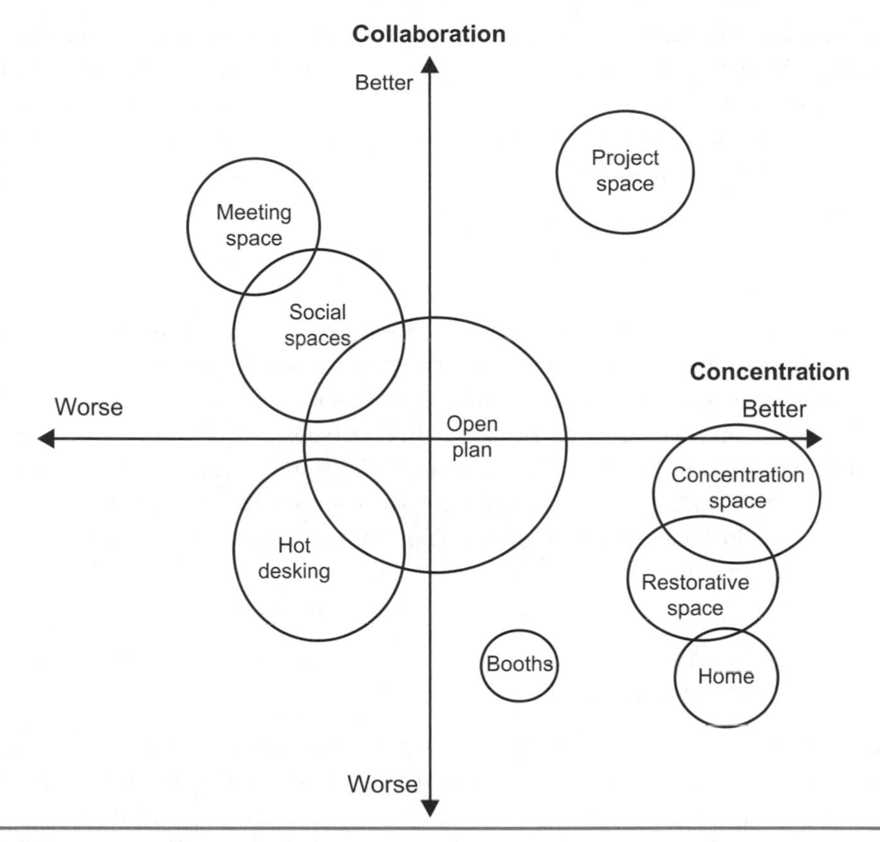

Figure 7.3 Future office workspaces for older knowledge workers. *(From Erlich & Bichard, 2008, p. 282; reproduced by permission of Emerald)*

TABLE 7.1 Knowledge Work Equals Four Work Modes (Gensler, 2008)

Collaborate	Focus	Learn	Socialize
Innovative Capital	*Productive Capital*	*Intellectual Capital*	*Social Capital*
Working with another person or group to achieve a goal	Work involving concentration and attention to a particular task or project	Working to acquire new knowledge of a subject or skill through education or experience	Work interactions that create common bonds and values, collective identity, collegiality, and productive relationships

- **Socialize.** This work mode identifies the benefits of social interaction in the office environment. It acknowledges the benefits of the office social network as a way of achieving enhanced collaboration.

Steelcase, the international office furniture manufacturer, identify four different work modes which are based on the way that people work and communicate within their office environment (Steelcase, 2009). The four different work modes are fundamentally based on the number of people involved in the work process. The four work modes are summarized below:

- **Working on your own (one person).** The work settings created to support the individual worker could range from a touchdown base for a mobile worker to a dedicated desk for a location-required worker.
- **Working in pairs (two persons).** This work entails collaboration between two people. However, it is acknowledged that the work setting must allow both individual work and work involving one other person.
- **Working in small groups (three to five persons).** This work mode could be considered to be team based working. The work setting created will need to support the team's need to collaborate, share knowledge and transfer learning.
- **Working in large groups (greater than six persons).** The work setting that supports this mode may include meeting spaces or project spaces. The spaces created should assist in sharing and communicating information with others.

In their book *Space to work,* Myerson & Ross (2006) propose four realms of knowledge work: Academy, Guild, Agora and Lodge:

- **Academy** – sees the organization as a learning organization. The workstyle adopted will be collaborative and collegiate. This approach ensures that knowledge is easily shared between colleagues. The term 'learning campus' could be used to encapsulate the principles of this working environment. This realm can be termed the corporate realm.
- **Guild** – encompasses the professional relationships the knowledge worker may have with their peers. In this realm the knowledge worker meets with their professional peers with the purpose of sharing ideas, skills and knowledge. The relationships would probably be developed outside the workplace and in an institution or association accommodation. This realm can be termed the professional realm.
- **Agora** – captures the relationship an organization has with its customers. The organization is open to the markets, customers and other sectors. The interactions could be undertaken in public spaces with the aim of bringing the organization closer to its customers and markets. This realm can be termed the public realm.
- **Lodge** – describes the home work setting where work can be undertaken when not in the office environment. This may address the issues of work–life balance as long as the boundaries of work and family life are established. This realm can be termed the domestic or private realm.

The challenge facing the knowledge worker is the balancing of the potential conflicting sets of relationships in the four different realms. The four relationships include:

- colleagues in the employing organization;
- professional peers;
- customers in the marketplace;
- family and friends in the home.

An essential ingredient that could allow the knowledge worker to maintain and service the different sets of relationships would be mobility. Since each of the four realms' activities are undertaken in different locations, the knowledge worker needs to have the freedom to move from location to location as the work demands require.

A publication produced in 2008 by DEGW and the Office of Government Commerce (OGC), entitled *Working beyond walls: the government workplace as an agent of change* attempts to categorize new work styles with the aim of creating an understanding of how and where people actually work. Three main work style categories are proposed, with each category having two types of work style (Hardy et al., 2008). The work style categories are set out below:

- **Residents.** This type of work style requires the worker to be largely working within the office environment. This is a style of working that could be classified as office-location-required. Resident workers will tend to be either team anchors or process workers. A resident worker will have their own office desk.
- **Internally mobile.** Internally mobile workers have more flexibility in how they work in the office environment. This may mean they use a range of different locations within the office or the building. There is less of a requirement for a dedicated office desk as the internally mobile workers will more likely use shared office desks. However, there will be an increase in need for mobile ICT systems. Internally mobile workers will tend to be knowledge workers or managers.
- **Externally mobile.** Externally mobile workers will tend to work largely outside the office environment. The work characteristics adopted will be either nomad workers or home/remote workers. This means that they will not require a dedicated desk but will use shared office desks when actually in the office environment. This could be classed as the ultimate in non-territorial working.

In developing a workplace strategy, establishing a clear understanding of where the work takes place can be just as important as establishing the type of work undertaken. A key ingredient in establishing the success of any future strategy would be the matching of the right type of space to the actual work requirement. The Nokia case study illustrates how the office environment can be designed to accommodate mobile workers.

The main data gathering and analytical techniques required to undertake work style analysis are summarized below.

- Site survey. The current working environment would be evaluated to establish current working practices and current space standards.
- Interviews. This would include representatives from the office environments. The interviews would aim to establish the work processes adopted by the office occupiers.
- Observation. A possible technique that could be used to monitor actual physical interactions between departments/functions and the movements around a building.
- Cultural readiness to change. The data gathering technique adopted would be a survey with the aim of establishing office occupiers' readiness to change their working practices.
- Occupier surveys. These would take the form of a questionnaire and would aim to establish the office occupiers' satisfaction with their current working environment. Included in the survey would be the identification of factors that the office occupiers felt were important in their office environment.
- Utilization study. A monitoring technique that establishes how current space and working environments are utilized.

Procedures and Protocols

As previously identified, the office space in an organization can send clear messages about the organization's brand and identity. It also gives physical cues to the organizational culture and sends signals about *'what it is like to work around here'*. Establishing the way an office space is to be used requires the creation of office protocols. Identifying office protocols requires a thorough understanding of the social dynamics of an office environment. The behavioural aspects of the office environment require protocols to ensure that distraction and interruptions do not impact on office occupiers' productivity.

When developing new office protocols and procedures it is often useful to establish a checklist (Hardy et al., 2008). Some of the main considerations are set out below.

- **Noise.** Conversations in an office environment may be a sign of interaction and collaboration but they may also be a form of distraction and irritation to someone who is not involved in the conversation. Protocols should be developed to establish how conversations are to be managed in an open plan environment. This could include the creation of small meeting areas for ad-hoc conversations. Another type of noise is the constant ringing of a telephone when people are away from the desk. This can be addressed by people putting their telephones onto voice mail when they are not at their desk.

- **Distraction-free work.** This requires individual concentrated work to be respected and valued. When working in an open office environment people have to acquire a certain amount of sensitivity to other people's workstyle.
- **Security.** Ensure that computer log-in details and passwords are kept safe. In addition, people should lock up their valuables in their desks or lockers. Visitors to the organization should be escorted and have a visible visitor pass.
- **Clutter.** Ensuring people's desks and work areas are tidy can give an office environment a general feeling of orderliness, but it can also be an essential office procedure if a shared desk policy is adopted.
- **Food.** People tend to eat their lunch whilst sat at their desk and if they eat food that has a strong odour it can permeate the office for the remainder of the day. Protocols should be introduced to give people guidance as to where and when it is appropriate to eat food in the office.

Having established the appropriate office protocols and procedures it is important that they are maintained (Hardy et al., 2008). It is essential that the way the office space is used and the way that people behave in that space is monitored and managed. If office protocols are violated then the team leader or office manager needs to be informed so that they can take the appropriate action.

Briefing Process

An essential component that links organizational demand to accommodation and work space provision is the briefing process. The briefing process aims to transfer the potentially abstract concepts of the organizational requirements into detailed building specifications. The Commission for Architecture and the Built Environment (CABE) identifies a number of phases to the briefing process (CABE, 2003), which are summarized below.

- **Vision statement.** This part of the process aims to capture the organization's expectations and visions of what a new workplace would look and feel like.
- **Outline brief.** This document aims to transfer some of the abstract organizational requirements into more of a statement of organizational needs. This document is sometimes referred to as the strategic brief (Marmot, Eley & Bradley, 2005).
- **Detailed brief.** This part of the briefing process moves the project further into the design phase. Through an iterative process of client consultation a more detailed understanding of organizational demand is identified.

- **Specification.** It is at this stage of the briefing process that the full details of building design are specified.
- **Building manual.** This document is usually provided to the end user on compilation of the building. It acts as an operation and maintenance manual for the efficient and effective use of the building.

Strategic Brief

As previously identified, this document is the linking document between the abstract vision statement and the detailed project brief (Marmot, Eley & Bradley, 2005). This is an essential part of the briefing process as it requires the CREAM manager to clearly establish and understand the needs of the organization, sometimes termed the strategic intent. The strategic intent is then converted into a demand profile that can be developed and used to evaluate the building supply.

An integral component of the strategic brief is the space budget. The space budget establishes the Net Usable Area (NUA) requirements of the organization. This is the space an organization actually uses in a building. The exact terms and phrases used to explain the NUA may vary, however NUA tends to consist of the following elements (Eley & Marmot, 1995; McGregor & Then, 1999):

- **Workspace.** This is space for individual work area requirements. It will include space for a desk and personal filing. This space can also include the local circulation or secondary circulation space required to reach the workspace.
- **Ancillary.** This is space allocated to people working as a group or department. It will include such areas as: storage, filing, local copying, meeting room, project area, interview room and vending machines.
- **Support.** This space is general organizational space. It will include such areas as: boardroom, library, central archive, central printing, gym, canteen and reception area.
- **Fit factor.** This is a contingency as buildings rarely fit the exact needs of organizations.

The main purpose of the strategic brief document is to meet the future business needs of the organization by ensuring the appropriate provision of workspace. The strategic brief may include alternative workplace scenarios which could form the basis for informed decision-making.

Building and Workplace Supply

Once the organizational demand has been defined, it is the CREAM manager's responsibility to identify the most appropriate building and workplace provision. The choice of building and workplace supply is of significant strategic importance to an organization and can have a direct impact on the future performance of the business.

It is at this point in the relocation process that the CREAM manager translates organizational demand for space into criteria that can be used to assess building supply. It is essential that the CREAM manager removes any subjectivity from the decision making process if any potential mismatch between organizational demand and building supply is to be avoided.

The size of the organization can have a significant impact on the building evaluation and decision making process (Greenhalgh, 2008). Research undertaken by Wrigglesworth & Nunnington (2004) suggests that larger firms are more likely to pursue a sophisticated measurement and modelling process. In addition, smaller firms with less resources were more likely to make the relocation decision based on 'gut feeling' rather than detailed evaluation (Wrigglesworth & Nunnington, 2004). The research also confirmed that some decisions were made in a few cases based on personal preferences of the Chairman or Chief Executive. However, we believe that with increased transparency, accountability and corporate social responsibility, decisions based on more rigorous and objective approaches will be demanded.

Establishing clear quantifiable building assessment criteria and a framework for evaluation are essential if the relocation decision is to be made in a purely objective manner. The methodology most likely to be adopted would be one that compares the specifications for building demand with the specifications for building supply. However, Wigglesworth & Nunnington (2004) propose that the relocation evaluation process should include more than just the evaluation of the building specification.

> *The creation of checklists and tick box exercises is common. For companies with a wider brief, another set of factors such as demographics and quality of life are likely to play a greater role.*
>
> **Wrigglesworth & Nunnington, 2004, p. 2**

We will now explore the key stages in the building evaluation decision making process. We will demonstrate how the organizational demand for space can be translated into a building supply profile which can be subsequently used to evaluate building supply possibilities.

The supply profile breaks down into a number of main components:

- the macro location (country) where an international search is being made;
- the macro location (city) where a national search is being made;
- the micro location (options within selected city);
- the micro location (characteristics of selected location);
- the building specification;
- the building configuration;
- specific operational requirements and prerequisites.

Where organizations are examining a range of potential locations, **macro** factors considered most important according to the Cushman & Wakefield European Cities Monitor 2008 are set out in Table 7.2:

Of course each organization will have its own ranking of priorities dictated by the nature of its business; organizational goals and aims of the relocation. This in turn will dictate the final selection of Country and City location.

At the more **micro** level, factors to be considered will include:

- accessibility to motorway;
- traffic flow - congestion;
- access to main railway station, bus and tram services;
- public security – lighting etc.;
- proximity to hotel accommodation;
- proximity to shops/services – e.g. cash dispenser/dry cleaning;
- proximity to restaurants/coffee shops/cafes;
- parking standards – on-site;
- distance to public parking;
- security;
- infrastructure – gas/electricity/alternative energy sources.

TABLE 7.2 Extract from Cushman & Wakefield European Cities Monitor, 2008

Factor	2008 %	2007 %
Availability of qualified staff	60	62
Easy access to markets, customers or clients	59	58
The quality of telecommunications	54	55
Transport links with other cities and internationally	53	52
Cost of staff	40	36
The climate governments create for business through tax policies or financial incentives	27	27
Languages spoken	27	29
Value for money of office space	26	26
Ease of travelling around within the city	25	24
Availability of office space	24	26
The quality of life for employees	21	21
Freedom from pollution	18	16

This survey, researched independently for Cushman & Wakefield by TNS (Taylor Nelson Sofres), engaged with senior executives from 500 European companies. Only 'Absolutely essential' responses are included in this table.

At the building level, factors to be considered may include:

- BREEAM rating;
- EPC rating;
- lighting specification;
- controllability of air handling;
- lift capacity – average wait time at peak periods;
- reception;
- access control;
- toilet capacity – number of employees per cubicles per floor;
- solar shading;
- fire safety system;
- access of daylight: distance from furthest work area to an elevation;
- horizontal flexibility – structural planning grid;
- vertical flexibility – floor to ceiling height;
- building reflects desired company image;
- 24/7 operational availability;
- broadband – communications infrastructure;
- ease of maintenance and cleaning of building;
- showers (for cyclists) etc.

Again the weighting placed on each element will be individual to the organization. What we find is that extracting the preferences and ranking them in an objective way is difficult for companies and they need time, support and a simple framework to assist them in their supply prioritization.

We have been working for many years with the Dutch designed system the Real Estate Norm (REN), originally developed by DTZ Zadelhoff and Jones Lang Wootton (now Jones Lang LaSalle) and G&P Starke Diekstra. It has never really been adopted fully in the industry, probably due to, in our opinion, to its overcomplexity and attention to detail and Dutch orientation. However, we believe that its methodology in identifying a comprehensive list of occupier demand factors, translating this into a specification and allowing users to score available buildings, is a very useful approach. The authors have been working with Creativesheffield, the Sheffield based urban regeneration company, to extend, refine and adapt the REN approach into a working tool to support the organization demand and supply analysis, inherent in relocation projects.

The tool is spreadsheet based and requires occupiers to rank 54 business factors in terms of High, Medium, Low priority or Not Applicable. Each of these factors has been examined in detail and a building specification developed on a 5-point scale. For example, the factor **Access to Main Railway Station** is defined on a 5-point scale as:

1 (Highest)	<5 minutes walk
2	<10 minutes walk
3	<15 minutes walk
4	15-30 minutes walk
5 (Lowest)	>30 minutes walk

This allows the organization and/or its consultants to consider each factor in terms of both demand (business priority) and specification. So, for example, if an organization is wishing to be sustainable and maximize its employees' access to public transport and it believes it may draw staff regionally, it may make access to a main railway station a high priority and score the factor with a 1, that is, it requires a main railway station within 5 minutes walk of its facility.

The Creativesheffield model starts with the assumption that Sheffield has been selected as the, or one of the, cities for relocation. It then divides the criteria into five sections:

- macro location within Sheffield;
- micro location – the immediate surrounding area of the building;
- building specification;
- building configuration;
- operational and other requirements.

Figure 7.4 illustrates one section of the Creativesheffield tool (Macro Location).

When a client has worked through the spreadsheet confirming the business priorities and desired specifications for each component the tool can then be used to score a shortlist of potential buildings by assessing what they actually provide on the 1–5 scale. So, returning to our main railway station criterion, we may find some buildings that meet the client's requirements of less than 5 minutes (scoring a 1) and others that do not.

The system can then be used in the same way as REN to create a matching profile, or deviation chart showing how the shortlisted buildings meet the client's requirements. This can be done in a number of ways, for example using a weighted score with the high priority elements being weighted highest to derive an overall matching score for each building.

However we prefer a chart-based approach that acts as a kind of fingerprint for each building and clearly demonstrates which building is most suitable against the client's priorities. As an illustration, if we consider just 5 of the 54 criteria in the

Creativesheffield

Sheffield Hallam University
SHARPENS YOUR THINKING

Objective Business Criteria to Real Estate Matching Tool

Client Demand Profile	Business Priority				High ----- Requirement/Specification				----- Low
	High	Medium	Low	N/A	1	2	3	4	5
Section 1 MACRO LOCATION within SHEFFIELD									
1 Location					Central CBD	Secondary Central	Edge of City Centre	Suburban	Rural
2 Type of Surrounding Environment					Central CBD	Suburban	Industrial	Residential	Peripheral
3 Accessibility to Motorway					<5 minutes drive	<15 minutes drive	<30 minutes drive	30-45 minutes drive	>45 minutes drive
4 Traffic Flow - congestion					100% Clear	Limited at Peak times	Constant at Peak times	During Office Hours	Constant
5 Access to Main Railway Station					<5 minutes walk	<10 minutes walk	<15 minutes walk	15-30 minutes walk	>30 minutes walk
6 Public security - lighting etc.					Fully lit and CCTV	Fully Lit	Partially Lit	Poorly Lit	None
7 Proximity to Hotel Accommodation - at least 2 within					<5 minutes walk	<10 minutes walk	<15 minutes walk	15-30 minutes walk	>30 minutes walk
8 Land Tenure					Freehold - unrestricted	Long LH>99 years	Long LH<99 years	Occ. LH 1<25 years	Licence

Figure 7.4 Illustrative component of the Creativesheffield Business Criteria to Real Estate Matching Tool developed by the authors

	⇐ Over-specification			Building A	Under-specification ⇒		
Macro location factor				**Client's required specification**			
Accessibility to motorway			1	2			
Traffic flow *congestion*				3			
Access to main railway station				1			4
Public security *lighting etc.*			1	2			
Proximity to hotel accommodation				2		4	

Figure 7.5 Illustrative example of a deviation chart for a single building of the Creativesheffield Business Criteria to Real Estate Matching Tool, developed by the authors

Creativesheffield tool we can create a chart as shown in Figure 7.5, which looks at how well a shortlisted building matches the client's priorities. The chart shows the client's required specification (1–5) in the central column of the deviation chart. Table 7.3 explains the deviation scoring for each macro location factor.

A negative score, where the building has an under-specification, is always seen as more of a problem than an over-specification and will be highlighted in red on the actual charts created by the system. This is because it is usually much more difficult to deal with an under-specification. For example, if a client requires full air conditioning it may be impossible to retrofit this into certain buildings. Whereas, if the client requires comfort cooling but looks at a building with full air conditioning there may be a financial penalty but no major technical barriers. Building A in the illustration is unlikely to meet the needs of the company, especially if the highlighted factors are ranked as a high priority.

TABLE 7.3 Demand to Supply Deviation Scoring Explanations

Macro location factor	Deviation score and explanation
Access to main railway station (Client Demands = 1)	Building A supplies a 4 score in its specification and therefore as the requirement was for 1, a negative deviation of 3 is shown as three bars to the right
Accessibility to motorway; and public security (Client Demand = 2)	Building A supplies a 1 (highest) score in its specification and therefore as the requirement was for 2 a positive deviation of 1 is shown as one bar to the left
Proximity to hotel accommodation (Client Demand = 2)	Building A supplies a 4 score in its specification and therefore as the requirement was for 2 a negative deviation of 2 is shown as two bars to the right
Traffic flow (congestion) (Client Demand = 3)	Building A supplies a 3 score and therefore the demand and supply are equal so there is no deviation

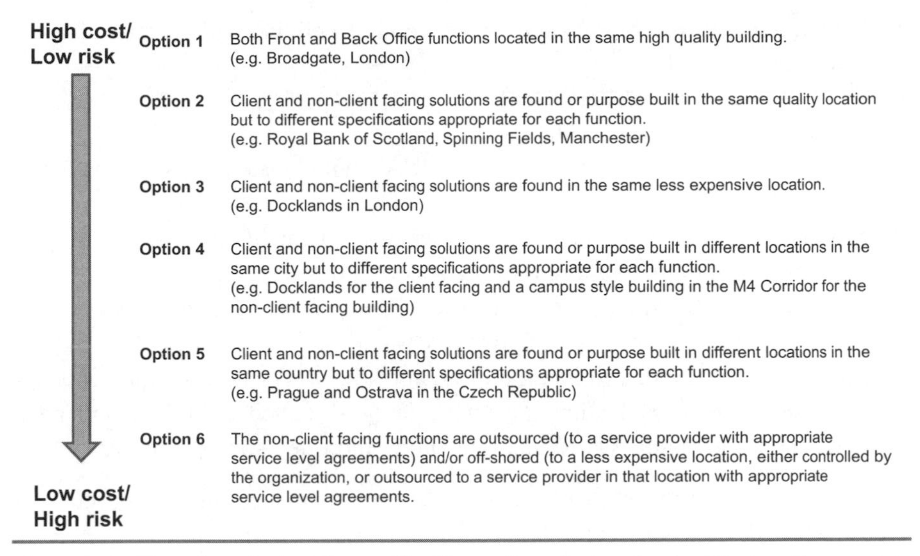

Figure 7.6 The client facing (front office)/non-client facing (back office) options matrix

What this tool facilitates is an objective, robust matching process that allows easy comparison of buildings and their ability to deliver against the client's key business requirements. The tool does, of course, examine all of the 54 criteria appropriate to a particular client to give a comprehensive, objective but manageable analysis of available building supply.

The relocation decision may also involve the consideration of both client facing (front office) and non-client facing (back office) activities at the same time. There are a number of different options available to organizations that should be considered in terms of the costs and risks associated with each option. These are set out in Figure 7.6.

Optimum Alignment

A corporate relocation is an opportune time to ensure that the building and workplace supply match the organizational demand. In Chapter 2 we proposed a CREAM strategic alignment model which illustrated how the matching of CREAM provision with organizational demand could lead to enhanced organizational performance. If the optimal alignment of organizational demand and CREAM provision is not achieved then this can ultimately lead to a potential mismatch of the building and workplace provision (McGregor & Then, 1999). Achieving optimal alignment of CREAM requires detailed evaluation, as set out below.

- **Building location.** The choice of the location of building will be determined by macro criteria such as:
  - close to customers/clients; and
  - located near appropriate labour pool.
- **Space provision.** The Net Usable Area (NUA) should match the NUA demand of the organization.
- **Building specification.** The building services in the building should be of sufficient specification to avoid retrofitting at a later date.
- **Floor plate of building.** The floor plate of the building should allow optimum layout for the different types of workspaces.
- **Space layout.** The design and the location of the workspaces should match the demands of the organization. The space layouts created should support inter-department interaction whilst also supporting intra-department work processes.

An essential component that is integral to the optimum alignment approach is the management of the change process.

Change Management

Relocation of a head office building includes both organizational change and individual change as experienced by the employees. Changing an employee's working environment can evoke an emotional response. It is important to recognize the links or attachments that employees create with their workplace (Inalhan, 2009).

Establishing the employees' *'readiness to change'* can help in identifying the people most likely to have difficulty in the change process. During a change management process there are likely to be three distinct categories of people (Laframboise, Nelson & Schmaltz, 2003).

- The first category could include people who may be initially surprised by the proposed changes but could ultimately be influenced to accept the proposed changes.
- The second category could include people who are excited by the change prospect and are generally supportive of the changes proposed.
- The final category could include people who do not see the need for change and can ultimately be the resisters to the proposed changes.

One key strategy that can be adopted to ensure that all employees engage with the change management programme is a communication strategy. The communication tools and techniques adopted should be appropriate to the stages of the change management programme. Research undertaken by Laframboise, Nelson & Schmaltz (2003) identified the stages of change appropriate for a relocation project to be: discovery, denial, resistance and acceptance (Laframboise et al., 2003). Each of these stages will now be discussed with special emphasis placed on the appropriate communication tools (Laframboise et al., 2003), as set out below.

- **Discovery.** This is the stage at which an employee finds out about the proposed changes to their working environment. This is an important stage in the change management process and therefore must not be left to chance. How and when employees hear about the change should be planned for in the communication plan. A possible vehicle for communication could be a town hall meeting. This approach gives the senior management an opportunity to explain their vision for the future project.
- **Denial.** It is at this stage of the change process that people continue in a 'business as usual' way of working and may ignore the new working environment proposals. A communication tool that could be adopted at this stage could be a project room/wall, which includes the new working environment layouts and proposed timescales. This approach keeps reminding employees that a change to their working environment is going to take place. An additional supporting communication tool could be the creation of a project website which gives all the up-to-date information on the progress of the project.
- **Resistance.** During this stage of the change process employees start to acknowledge that their working environment is going to change. This is where employees may demonstrate emotions such as anger and resentment (Inalhan, 2009). It is essential that employees concerns are listened to and addressed appropriately. One of the major fears employees' experience is the inability to influence the changes that will have a direct impact on their working lives. Therefore, an appropriate strategy would be to involve employees in the decision making process. Adopting this approach should ensure that the work environment created actually links to the work processes undertaken by the employees.
- **Acceptance.** The last stage of the change management process is the acceptance stage. It is at this stage that the employees accept the new working environment. This is an opportune time to celebrate the successes of the project. A tool that can evaluate employee satisfaction with the new workplace is the post-occupancy evaluation (POE). This is usually undertaken once employees have been in their new working environment for approximately six months. The POE gives the employees an opportunity to feedback their satisfaction or dissatisfaction with the new working environment. An extended after-care and POE service consisting of one to three years could be adopted to ensure that employees get the most from their workspace and building (Way, Bordass, Leaman & Bunn, 2009).

A practical application of the linkage between communication and change management in a workplace project can be seen in the Nokia case study.

Summary Checklist

1. A way of establishing the organizational demand for space is through the use of a strategic brief. During the preparation of the brief consideration should be given to:
   - **a.** An evaluation of the organizational aims, objectives and mission statement.
   - **b.** Establishing a deeper understanding of the nature of the organization's business.
   - **c.** Evaluate the organizational chart, which should include current and forecast headcounts.
   - **d.** Identify the current organizational culture and the future cultural aspirations of the organization.
   - **e.** Establish funding for the relocation project and prepare estimated costings.
   - **f.** Identify the key relationships between departments using an adjacency diagram.
   - **g.** Establish the brand identity that the organization wishes to represent through the new office environment.
   - **h.** Evaluate the potential impact ICT could have on the accommodation requirements.
   - **i.** Undertake a work style analysis so that appropriate spaces can be created.
   - **j.** Create a space budget that will establish the net usable area (NUA) requirement for the organization.
   - **k.** Finally, introduce protocols and procedures to ensure the office space created by the relocation project is managed effectively.
2. Once the organizational demand has been established possible locations and buildings can be evaluated. The building and workplace appraisal should include an evaluation of:
   - **a.** The macro location (country) where an international search is being made.
   - **b.** The macro location (city) where a national search is being made.
   - **c.** The micro location (options within selected city).
   - **d.** The micro location (characteristics of selected location).
   - **e.** The building specification.
   - **f.** The building configuration.
   - **g.** Specific operational requirements and prerequisites.
3. Apply the CREAM strategic alignment model proposed in Chapter 2 to ensure that the building and workplace supply match the organizational demand.
4. Ensure that communication strategies and a change management programme are set up to run in conjunction with the relocation project.

References

CABE. (2003). *Creating excellent buildings: A guide for clients*. London: Commission for Achitecture and the Built Environment.

Cushman, & Wakefield. (2008). *Cushman & Wakefield European Cities Monitor*. http://www.cushwake.com/cwglobal/jsp/kcReportDetail.jsp?Country=GLOBAL&Language=EN&catId=100003&pId=c17500010p. Accessed 21.10.09.

Eley, J., & Marmot, A. (1995). *Understanding offices: What every manager needs to know about office buildings*. London: Penguin Books.

Erlich, A., & Bichard, J.-A. (2008). The Welcoming Workplace: designing for ageing knowledge workers. *Journal of Corporate Real Estate, 10*(4), 273–285.

Gensler. (2008). *2008 Workplace Survey: United Kingdom.* London: Gensler.

Greenhalgh, P. (2008). An examination of business occupier relocation decision making: distinguishing small and large firm behaviour. *Journal of Property Research, 25*(2), 107–126.

Hardy, B., Graham, R., Stansall, P., White, A., Harrison, A., Bell, A., et al. (2008). *Working beyond walls: The government workplace as an agent of change.* London: DEGW & OGC.

Inalhan, G. (2009). Attachments: the unrecognised link between employees and their workplace (in change management projects). *Journal of Corporate Real Estate, 11*(1), 17–37.

Johnson, G., & Scholes, K. (1999). *Exploring corporate strategy text and cases* (5th ed.). Harlow: Prentice–Hall.

Laframboise, D., Nelson, R., & Schmaltz, J. (2003). Managing resistance to change in workplace accomodation projects. *Journal of Facilities Management, 1*(4), 306–321.

Marmot, A., Eley, J., & Bradley, S. (2005). Programming/briefing – programme review. In W. Preiser & J. Vischer (Eds.), *Assessing building performance* (pp. 39–51). Oxford: Elsevier Butterworth–Heinemann.

McGregor, W., & Then, D. (1999). *Facilities management and the business of space.* New York, NY: Arnold.

Myerson, J., & Ross, P. (2006). *Space to work.* London: Laurence King Publishing.

Schramm, U. (2005). Strategic planning - effectiveness review. In W. Preiser, & J. C. Vischer (Eds.), *Assessing building performance* (pp. 29–38). Oxford: Elsevier Butterworth–Heinemann.

Smith, J. (2008). *Welcoming Workplace: Designing office space for an ageing workforce in the 21st century knowledge economy.* London: Helen Hamlyn Centre, Royal College of Art.

Steelcase. (2009). *Steelcase space planning.* http://www.steelcase.com/eu/steelcase_space_planning_news.aspx?f=36808. Accessed 21.07.09.

Way, M., Bordass, B., Leaman, A., & Bunn, R. (2009). *The SOFT LANDINGS FRAMEWORK for better briefing, design, handover and building performance in-use.* Bracknell: BSRIA & UBT (Usable Buildings Trust).

Wrigglesworth, P., & Nunnington, N. (2004). *Reasons for relocation: Corporate property professionals' views.* Retrieved from RICS. www.rics.org/management/facilitiesmanagement/businesslocation/reasonsforrelocation. Accessed 04.09.06.

Chapter 8

Sustainability, Corporate Social Responsibility and Corporate Real Estate

Introduction

The built environment has a significant role to play in the sustainability agenda. The Organization for Economic Co-operation and Development (OECD), in its analysis of energy consumption of OECD countries (2005), found that buildings and infrastructure represented from 40 to 50% of gross capital formation for each of the countries, and between 25 and 40% of final energy consumption. A survey by Gensler (2008) reported that 73% of property developers believe that spiralling energy costs will make energy consumption a significant factor in building selection and drive the development of more energy-efficient buildings.

Sustainability is a topic growing rapidly in significance and there are many books that cover the topic in detail. In this chapter we wish to approach it from the perspective of a corporate occupier and examine the impacts of sustainability on corporate real estate. More and more firms are paying attention to sustainability and trying to reconcile their corporate social responsibility statements with their occupation of real estate. It is, for example, a highly visible contradiction to have sustainability as a high profile statement in your annual Corporate Social Responsibility (CSR) report and occupy a building that has an energy performance certificate on view next to the reception desk showing the building to be highly inefficient!

Definitions

Sustainability

> *A sustainable society is one which satisfies its needs without diminishing the prospect of future generations.*
>
> Lester R. Brown, Founder and President, Worldwatch Institute (Brown & Wolf, 1998)

Corporate Real Estate Asset Management. ISBN: 978-0-7282-0573-4

Leave the world better than you found it, take no more than you need, try not to harm life or the environment, make amends if you do.

Paul Hawken, 1994, p. 13

Sustainable development:

In essence, sustainable development refers to: 'development which meets the needs of the present without compromising the ability of future generations to meet their own needs'.

The Bruntland Commission, 1987

Sustainable buildings:

Sustainable buildings can be defined as 'those buildings that have minimum adverse impacts on the built and natural environment, in terms of the buildings themselves, their immediate surroundings and the broader regional and global setting'.

Beard & Roper, 2006, p. 4

Green buildings:

Green buildings are designed to reduce the overall impact of the built environment on human health and the natural environment by:

- Efficiently using energy, water, and other resources;
- Protecting occupant health and improving employee productivity; and
- Reducing waste, pollution and environmental degradation

US Environmental Protection Agency, 2009

Corporate Social Responsibility (CSR)

CSR does not have a universal definition; in fact it is sometimes quite difficult to find definitive statements about CSR. Many organizations regard CSR as the private sector's way of integrating the economic, social and environmental imperatives of their activities into their business operation. CSR is therefore very closely linked to the business pursuit of sustainable development and the USA concept of the triple bottom line. The triple bottom line may be attributed to the book of the same name by Andy Savitz (2006), former head of PricewaterhouseCooper's Sustainability Business Services, which outlines an innovative way to measure the bottom line – where profits go side-by-side with environmental and social performance.

In the UK, Business Link, a free business support service set up by the UK government, provides comprehensive resources and case studies on CSR on its website (www.businesslink.gov.uk). Under the section of its guide to CSR with the title *'the business benefits of corporate social responsibility'* it states that:

> *Corporate social responsibility (CSR) isn't just about doing the right thing.*

The guide also examines direct business benefits of CSR, which it states as:

- a good reputation makes it easier to recruit employees;
- employees stay longer, reducing the costs and disruption of recruitment and retraining;
- employees are better motivated and more productive;
- CSR helps ensure you comply with regulatory requirements;
- activities such as involvement with the local community are ideal opportunities to generate positive press coverage;
- good relationships with local authorities makes doing business easier;
- understanding the wider impact of your business can help you create profitable new products and services;
- CSR can make you more competitive and reduces the risk of sudden damage to your reputation (and sales). Investors recognize this and are more willing to finance you.

The guide examines how CSR sets business apart and draws customers who prefer to buy from an ethical business. It uses Ben & Jerry's as an excellent case study of how an integrated strategy focusing on both CSR and quality produce has led to a fast-growing premium priced business.

Interestingly, the guide blurs the boundaries between CSR and sustainability with a real estate example, where it states: 'reducing waste and emissions doesn't just help the environment – it saves you money too. It's not difficult to cut utility bills and waste disposal costs, bringing immediate cash benefits'. This is a useful guide for helping businesses reduce energy costs, water waste etc. and clearly demonstrates the linkages between the two concepts.

The Business Benefits of Corporate Social Responsibility (CSR)

It is important to recognize that CSR isn't just about doing the right thing. As the Business Link guide demonstrates, it also offers direct business benefits.

Building a reputation as a responsible business differentiates your business and, as with Ben & Jerry's or the Body Shop, may create added value and marketing capital. Many consumers prefer to buy from ethical businesses and increasingly prefer to invest in them. Companies often favour suppliers who demonstrate responsible policies as this helps them to minimize the risk of any damage to their own reputations. Companies increasingly want to demonstrate their CSR by occupying sustainable buildings and later in this chapter we describe a number of initiatives that create, classify and evidence sustainable building options.

Some customers do not just prefer to deal with responsible companies, but insist on it. For example, sales of *'environmentally friendly'* products continue to grow – and these products often sell at a premium price, Ben & Jerry's ice cream and hybrid cars such as the Toyota Prius being the most quoted examples.

Reducing waste and emissions can save an organization substantial costs. In a survey of key decision makers by Gensler in 2006, property directors believed that more than 20% of their current energy consumption could be saved if buildings were designed with energy efficiency in mind (up to £122 million per annum) and developers believed that 27% could be saved (equivalent to £155 million per annum). It is not difficult to cut utility bills and waste disposal costs, bringing immediate cash benefits.

Sustainability and Corporate Business Thinking

It is clear that sustainability is being embedded in the thinking of senior executives and reinforced by its inclusion on MBA programmes, including MIT where Peter Senge, leading guru and founder of the principles of the learning organization, has turned his attention to the subject. In his book, *The necessary revolution: how individuals and organizations are working together to create a sustainable world* (2008), he details the way companies are embracing the challenges and opportunities around sustainability. He makes an interesting statement in the book:

> *what's becoming clearer to companies is what happens to stuff after you make it. In Europe today, you make a car, you have to take it back at the end of its lifetime. That's the law. Same is true for a lot of consumer electronics. That's a fundamental shift in a business model. A few companies in Europe led the charge to design those regulations and they became world leaders of design for remanufacture and design for recycling. If you design the car right, you can have a lot of value in it even when it's no longer efficient to operate.*
>
> **Senge, 2008, p. 6**

There are important parallels here with the real estate industry, where sustainability will focus on life cycles of materials, renewability, recyclability and longevity.

Sustainability and CSR in Action

Bringing It All Together: How Siemens Use Sustainability in Real Estate as Evidence of CSR

The Siemens Annual Corporate Social Responsibility Report for 2007 includes significant evidence of sustainability measures related to its corporate estate:

Green Buildings

1. *For Siemens, energy efficiency represents a major business opportunity that promises significant rates of growth. Our innovations and technological advancements are helping to conserve energy and reduce greenhouse gas emissions all over the world. And at our company locations, effective building management is enabling us to achieve significant energy savings.*
2. *According to expert forecasts, the world's primary energy consumption is likely to double between 2000 and 2050. At the same time, the price of oil and prices of energy linked to oil are increasing – due to high demand as well as to a growing raw material shortage. The impact on the environment is considerable and the consequences for the management of commercial real estate are significant: The costs of heating and of technical operations are increasing continuously. Our task, therefore, is to manage our buildings' energy- and resource-efficiency.*

Sustainable Buildings

1. *Siemens Real Estate (SRE) is responsible for managing more than 3000 locations in 30 countries on five continents. This figure also includes all those small locations which, due to their small environmental footprints and comparatively low resource consumption, are not tracked in Siemens' company-wide environmental data recording system SESIS. As part of its GREEN BUILDING initiative for existing buildings, SRE has developed a natural resources management system (NRM) comprising strategic as well as operational modules. The NRM monitors energy and water consumption at the office locations. At the strategic level, its purpose is to create sustainable industrial and office buildings. At the operational level, it serves to identify possibilities for reducing resource consumption and to implement appropriate measures through modernization work on buildings.*
2. *Real estate managers on the ground are notified of simple and immediately effective means of reducing resource consumption, identified on the basis of our Resource Check Guidebook. These measures include turning down the heating at night and adjusting it in line with the hours in which buildings are used. The next step generally involves initiatives such as installing*

frequency converters; replacing components, e.g. pumps and control units; and installing fittings that save water. For the most part, measures like these quickly produce tangible successes, while more complex, structural work on buildings, such as the insulation of outer walls, has a positive effect in the longer term.

3. *Other elements of their GREEN BUILDING initiative include life cycle costs analyses carried out to take into account investment and utilization costs in the design of new buildings, and unified certification of major remediation and construction projects around the world conducted in the context of our GREEN BUILDING certification programme. A second SRE manual, the Sustainable Building Design Guidebook, covers a wide range of options – everything from choice of location to the use of renewable energy sources and efficient building management – that can enhance buildings' life cycle sustainability.*

Current Economy Measures

1. *SRE Italy completed energy checks at three locations. As a result, heating times were adjusted in line with the business hours, energy-saving lighting systems have been installed, and lighting has been improved. Taken together, these measures alone represent a potential saving of €134,000 a year.*
2. *At the 400,000 m^2 office and R&D facility in Munich-Perlach (Germany), annual savings of €250,000 were achieved and carbon emissions reduced by 1450 metric tons by bringing the heating, ventilation and air conditioning system technology up to date.*
3. *Optimization of Siemens' 23,000 m^2 Lindenplatz location in Hamburg so that it now consumes 10,000 cubic metres less water, 584,000 kilowatt-hours less heating energy, and 408,000 kilowatt-hours less electricity and emits 380 tons less carbon dioxide a year.*

(**Source:** *Siemens (2007).* Siemens Corporate Responsibility Reports. *Retrieved from http://w1.siemens.com/responsibility/report/07/en/management/env_protection/green_building.htm accessed 5 September 2009)*

The Professional Perspective

The RICS (Royal Institute of Chartered Surveyors) in the UK has been working hard to promote sustainability and the roles of its membership in providing sustainable solutions and has a Presidential Commission on Sustainability. In its 2007 publication *Surveying sustainability: A short guide for the property professional*, it examines the sustainability issues in respect of the property life cycle and regards the

TABLE 8.1 Significant Factors for the Sustainable Occupation and Use of Buildings

Property life cycle stage	Social	Environmental	Economic
Occupation and Use (including refurbishment)	A better quality of life, built to last, clean, working and friendly	Energy efficient operation, effective maintenance, occupier recycling schemes, greywater systems etc.	Use of local suppliers and contractors Increase in occupier productivity through sustainable facility management

factors shown in Table 8.1 as being significant for the Occupation and Use (including refurbishment) stage of that life cycle.

Sustainability and Commercial Property

There is no single, accepted definition of sustainability within the commercial property sector. However, a range of environmental tools have been developed to help organizations develop their own criteria and work towards greening their property portfolio, acquisition or tenancy in the context of their business goals, corporate values and organizational culture. These include sustainable development measurement tools including the Building Research Establishment Environmental Assessment Method (BREEAM) in the UK; Leadership in Energy and Environmental Design (LEED) in the United States; Green Star in Australia and New Zealand; energy performance certificates and green leases. We will examine and compare these initiatives later in this chapter.

A number of aspects of sustainability are becoming more common and visible throughout the commercial property sector and they are adopted due to a number of drivers that we consider later in this chapter.

The current 'hot topics' include:

- recognizing that the environmental impacts of the sector, including impacts on climate change and potable water, are significant;
- expectations of significant increases in energy costs and ways in which these can be reduced;
- accepting the link between real estate and workplace, including the proven relationship between healthy workplaces and increased workforce productivity;
- increasing awareness of, and expectations of, new employees entering the workforce, including Generation Y and the next generation, Z (see Chapter 1);
- awareness that business value can be created by a comprehensive approach to sustainability;
- recognizing that all stakeholders, internal and external, are expecting continuing improvements in economic, social and environmental performance.

Drivers of Sustainability and How They Impact upon Corporate Real Estate

Drivers range from national government and EU policy directives to consumer expectations; quality compliance systems; changing attitudes of new generations of employees and customers; and a change (precipitated by the 2008 credit crunch) in the relationships between landlords and tenants. These drivers require occupiers and landlords to reconsider their responsibilities to improve their environmental performance, particularly in relation to energy consumption.

Some of these reasons are provided below, focusing on bottom-line business reasons, which all aim to reduce carbon emissions from the built environment and improve environmental performance.

The Energy Performance of Buildings Directive (EPBD), introduced in the UK in 2006, has led to the amendment of the UK Building Regulations, which will require all new commercial buildings to produce 27% less carbon than previously allowed. Under the EU Directive, the government is introducing new measures for inspecting and rating energy efficiency of a building, which will need to be declared when the building is sold, let or refurbished. Further Articles from within the EPBD are likely to be implemented in the future which will have direct relevance to the commercial sector. All buildings will have to be compliant by 2009.

Recent research from the UK government, the Royal Institute of Chartered Surveyors and CoreNet Global suggest that green buildings may command both a higher rent and be of greater investment value.

Central and local government conditions require public sector occupiers to meet increasingly exacting targets, including the occupation of BREAM buildings rated 'good' and above, and to move towards an average Energy Performance Certificate (EPC) rating of C.

Better-performing buildings may provide better value for the tenant in terms of the rent, whilst offering more flexible leases.

Compulsory carbon trading may be introduced by the government, which may include, amongst others, larger occupiers and property owners who use greater than £250,000 energy per annum.

In the United States a number of initiatives over the years have imposed expectations, benchmarks and requirements on the managers of the public estate, including:

- Federal Leadership in High Performance and Sustainable Buildings Memorandum of Understanding (January 2006);
- Energy Policy Act of 2005 (July 2005), Public Law 109-5.

In an interview for our Case Study 1, Rob Yates, Customer Service Director of Bruntwood, identified sustainability as both the most significant risk and the most significant market opportunity for the future.

International Standards

BS7750 was the world's first specification for an environmental management system, published in June 1992. The system was used to describe the company's environmental management system, evaluate its performance and to define policy, practices, objectives and targets; and provides a catalyst for continuous improvement. It was developed from more generic quality management approaches and standards. BS7750 was replaced by the international standard BS EN ISO 14001 in 1996, which was updated in 2004. ISO have many other standards relating to the environment but ISO 14001 is regarded as the building block for companies to establish their environmental approach.

Compliance to the standard is voluntary for companies, and complements required compliance to statutory legislation and is now designed to be compatible with the European Community's Eco-Management & Audit Scheme (EMAS). Companies can apply for certification that demonstrates their compliance to the standard.

According to ISO (2009):

> *The intention of ISO 14001:2004 is to provide a framework for a holistic, strategic approach to the organization's environmental policy, plans and actions. It does not specify levels of environmental performance as this would require a specific EMS standard for each business. ISO 14001:2004 gives the generic requirements for an environmental management system. The underlying philosophy is that whatever the organization's activity, the requirements of an effective EMS are the same.*

This has the effect of establishing a common reference point for dialogue about environmental management issues between organizations and their customers, regulators, the public and other stakeholders.

Because ISO 14001:2004 does not lay down levels of environmental performance, the standard can to be implemented by a wide variety of organizations, whatever their current level of environmental maturity. However, a commitment to compliance with applicable environmental legislation and regulations is required, along with a commitment to continual improvement – for which the EMS provides the framework.

ISO (2009) sets out the standards as:

- providing guidelines on the elements of an environmental management system and its implementation, and discussing principal issues involved;
- specifying the requirements for an environmental management system.

Fulfilling these requirements demands objective evidence that can be audited to demonstrate that the environmental management system is operating effectively in conformity to the standard.

ISO (2009) also sets out what the organization believes can be achieved by implementing the standard:

- introduction of a tool that can be used to meet internal objectives;
- providing assurance to management that it is in control of the organizational processes and activities that have an impact on the environment;
- assuring employees that they are working for an environmentally responsible organization;
- providing assurance on environmental issues to external stakeholders – such as customers, the community and regulatory agencies;
- supporting compliance with environmental regulations;
- supporting the organization's claims and communication about its own environmental policies, plans and actions;
- providing a framework for demonstrating conformity via suppliers' declarations of conformity, assessment of conformity by an external stakeholder – such as a business client – and for certification of conformity by an independent certification body.

It can be seen that for corporate occupiers, having an EMS is a fundamental building block to support all other sustainability initiatives. It embeds and integrates sustainability within the organization and provides management routines, protocols and review mechanisms that support increased sustainability. It also provides an opportunity for external verification and constant monitoring and revalidation of a company's EMS. It should also support and make more manageable many other environmental initiatives by providing infrastructure, direction and commitment.

Occupying Sustainable Labelled Buildings

One way in which a corporate occupier can visibly demonstrate sustainability is by occupying a certified building. There are many certification systems around the world including those shown in Table 8.2:

The four methods vary significantly and Table 8.3 shows a comparison of the assessment criteria used in these systems, using material produced by King Sturge (McIntosh, 2009).

Whist Table 8.3 shows that there are clear differences in the scope of criteria assessed by each system, the other main difference is in the operation and application of the systems, with CASBEE being the most differentiated. CASBEE was the first system to really consider the sustainability post completion, although BREEAM now has the facility to do this.

There are also many other systems, including those developed by private sector organizations such as Arup's Sustainable Project Appraisal Routine (SPeAR®).

TABLE 8.2 Some World Sustainability Certification Systems

UK	BREEAM	The Building Research Establishment's Environmental Assessment Method
USA	LEED	The Green Building Council's Leadership in Environmental and Energy Design
Japan	CASBEE	Comprehensive Assessment System for Building Environmental Efficiency
Australia/New Zealand	Green Star	Green Building Council of Australia

Whilst there has been some convergence of the systems over time they vary in the way they measure performance, the relative weightings on performance and the way in which the results are analysed and documented, making international comparability problematic.

There is insufficient space in this chapter to review all of the systems so we have focused on BREEAM, which is not only used in the UK but also across Europe and many of the BREEAM tools now have International versions.

TABLE 8.3 A Comparison of Some Certification Systems

Assessment criteria	BREEAM	LEED	CASBEE	Green Star
Energy	☑	☑	☑	☑
CO_2	☑			
Ecology	☑	☑	☑	☑
Economy				
Health and wellbeing	☑		☑	☑
Indoor environmental quality	☑	☑	☑	☑
Innovation	☑		☑	☑
Land use	☑	☑		
Management	☑		☑	☑
Materials	☑		☑	☑
Pollution	☑	☑	☑	☑
Renewable Technologies	☑	☑		☑
Transport	☑	☑		☑
Waste	☑			
Water	☑	☑	☑	☑

BREEAM

Although BREEAM is currently a voluntary scheme, achieving a BREEAM rating is increasingly becoming a requirement for public sector organizations, and future government directives could well make it a condition of occupation of and relocations to commercial buildings. For private sector organizations, both occupying a high-rated BREEAM building or undertaking an in-use assessment could pre-empt potential future legislative compliance, saving the organization time and money in the long term.

In addition to pre-empting future (probably inevitable) regulations, there are potential tangible business benefits too:

- staff retention and attraction;
- reduced utility bills;
- an improved internal environment;
- marketing and PR opportunities;
- increased productivity.

How does BREEAM Work?

A BREEAM assessment focuses on the following key areas:

- the impact a building has on the health and well-being of employees;
- the measures in place to reduce water usage and increase efficiency;
- the use of locally sourced materials, and responsible waste management both of construction and when the building is in use;
- the management policies in place, and how environmentally sound they are;
- the location of the building and its proximity to public transport, how it affects and manages transportation of people to and from work, and in some cases materials or products and the resulting impact this has on CO_2 emissions;
- building management operations including the integration of the 'Considerate Contractor' scheme, whereby contractors carry out their operations in a safe and considerate manner;
- the use of energy efficient measures and associated reduction in CO_2 emissions; in addition to Land Use, Site Ecological Value, and Pollution.

At the end of the assessment the building is given one of five ratings based on its environmental performance: Pass, Good, Very Good, Excellent and Outstanding. A building's BREEAM rating is calculated by adding together points awarded for a range of different sustainability criteria. Buildings with a score of more than 70 out of 100 are given the 'Excellent' rating, a score of 55–70 wins 'Very good' and so on. An appropriate BREEAM Certificate is awarded at the end of the process. BREEAM continues to evolve and the 2008 update represents the biggest single change that has happened to the scheme since its launch 18 years ago.

The key changes are:

- compulsory post-construction reviews mean that the completed building has to deliver on its design promises;
- minimum standards mean that high performance has to be achieved in key sections, particularly energy.

This addresses some criticisms of previous versions of BREEAM and brings it more in line with, for example, CASBEE. The new version of BREEAM, launched on 1 August 2008, makes a good score on energy efficiency and water consumption compulsory, not optional. The energy-efficiency element has also been raised from 15% to 19%. Buildings now have to improve significantly upon UK building regulations to score well on energy efficiency. It has been suggested that many buildings originally classified as 'Excellent' buildings would not meet this standard with the new energy-based requirements.

The changes reflect the way the environmental debate has evolved since BREEAM's launch over 10 years ago. Energy efficiency has become a higher priority as the link between carbon emissions and global warming has become more widely understood. The revisions will also ensure much better correlation between a high BREEAM score and good performance under energy performance certificates.

Whilst most people consider BREEAM to be an office rating system it is now being extended to include Retail, Healthcare, Hotels and Education buildings. Some examples of the diverse range of buildings which have achieved BREEAM Excellent status include:

Campus M, Munich	Developer: AIG
New Central Library, Cardiff	Developer: St David's Partnership
Central St Giles, London	Developer: Stanhope
Centrum Galerie, Dresden	Developer: Multi Development Germany
Asda Store, Bootle	Developer: Bowmer and Kirkland

BREEAM in Action

Digital Realty Trust

As reported in *FM Link*, April 2009, Digital Realty Trust has adopted the BREEAM Datacentre format as a basis for the design of its buildings throughout the world, anticipating significant environmental and cost savings.

> *We have created the new scheme using the BREEAM Bespoke methodology as the development framework to take account of the rise in interest in applying BREEAM to more specialist buildings. By working with industry leaders like Digital Realty Trust we are able to ensure that these new standards incorporate 'real world' experiences. The advantage of using BREEAM Bespoke for this is that it reduces the cost of assessments and provides a ready-made set of criteria against which developers, designers, and clients can assess their buildings. We will be refining the Datacentre scheme over the first 12 months of its operation following feedback from actual assessments, and in time these specialist schemes will become part of the standard suite for BREEAM.*
>
> **Paul Gibbon, Director of Sustainability at BRE Global, 2009**

Asda Store, Bootle, Merseyside

The £27 million ASDA store in Bootle achieved a score of 70.23% under BREEAM Retail 2006, qualifying for the 'Excellent' rating and ASDA's target of emitting 50% less carbon than a typical store and diverting 95% of operational waste away from landfill sites.

These are some of the features helped to make up the BREEAM score:

- 95% of operational waste is diverted from landfill through segregation and recycling;
- introduction of self closing mechanisms to cold storage cabinets, strip curtains and low energy lighting;
- waste heat recovery;
- extensive integration of natural lighting, including sun pipes and north facing roof lights;
- water efficient products, including passive infra red sensor urinals and low flow showers;
- a sedum green roof over the store's warehouse providing better water retention, insulation and a natural habitat;
- landscaping from locally sourced varieties that require minimal watering.

BREEAM and Its 'In Use' System

In addition to the main new build classification system, BREEAM is introducing an 'In Use' system. This enables building managers to self-assess the performance of their existing portfolio using the online BREEAM In-Use tool. Training is provided to license BREEAM In-Use Auditors to verify the building manager's self-assessment and provide a certificate.

BREEAM In-Use assessment differs from the established BREEAM schemes in that BREEAM assessors undertake an assessment of new buildings and submit the data to BRE Global to review and issue a certificate. The assessment tool will enable corporate real estate managers to see the impact of their building and existing systems and initiatives, as well as the potential impact of any proposed changes on the BREAAM points score and rating classification. This will enable corporate occupiers and their buildings managers potentially to:

- provide credible evidence of proven sustainability;
- provide a route to compliance with environmental legislation and standards, such as energy labelling and ISO 14001;
- reduce operational costs;
- enhance the value and marketability of property assets;
- give a transparent platform for negotiating building improvements with landlords and owners;
- encourage greater engagement with staff in implementing sustainable business practices;
- provide opportunities to improve staff satisfaction with the working environment with the potential for significant improvements in productivity;
- demonstrate and evidence an organization's commitment to CSR;
- improve organizational effectiveness.

Energy Performance Certificates

Directive 2002/91/EC of the European Parliament and Council, on the energy performance of buildings, came into force on 4 January 2003. Article 7 of the Directive dictates that all member states should ensure that:

- when a building is constructed, sold or rented, an energy performance certificate (EPC) is made available;
- an energy performance certificate should be placed in a prominent position and clearly visible to the public in buildings with a total useful floor area over 1000 m^2 occupied (in whole or in part) by public authorities and by institutions providing public services to a large number of persons.

The Directive states:

More energy efficient buildings provide better living conditions and save money for all citizens. The best moment for energy improvements is when buildings are constructed or are renovated. Member states have to ensure a good quality of the certificates and inspection processes.

Many EU countries initially failed to translate the Directive into national law by the initial deadline of 4 January 2006, and had to utilize the additional three-year extension to 4 January 2009. Many countries have still not fully implemented the Directive, with the inspection of boilers and air conditioning systems proving particularly difficult. Implementation of the initiative has also been patchy, with many small and medium landlords and occupiers not being aware of the implications of the Directive.

What Does an EPC Look Like?

Figure 8.1 shows an EPC displayed in a Sheffield Hallam University building.

A further resource to assist landlords and tenants in improving the energy efficiency of commercial buildings is available at www.LES-TER.org. The Landlords Energy Statement (LES) is a free tool which helps the landlord establish the energy/CO_2 requirement of providing communal services in the buildings and helps compare them against similar buildings with similar uses. It also identifies areas for improvements, and can be used to illustrate year-on-year improvements. The Tenants Energy Review (TER), which will be available on the site soon, helps tenants to quantify their direct energy use and identify occupancy features that influence energy demand. It also helps the tenant identify opportunities for improvement, sometimes in co-operation with the landlord. The TER in conjunction with the LES also helps collate the information required for an EPC.

The Measurability of Sustainability

Sustainability and corporate real estate has a mixture of 'hard' and 'soft' components, some of which are easy to measure, others are more problematic. The biggest challenge is for the industry to take an holistic view and link sustainability measures together (this is explored in Case Study 5). We have set out below examples of sustainability initiatives in a hierarchy of how easy they are to measure:

'Hard'/Tangible/Easily Measurable Aspects of Sustainability

- Energy costs measured before and after the introduction of energy saving initiatives such as motion sensors, LED light bulbs to replace halogen or solar glare protection.
- Water consumption measured on a before and after basis after the introduction of 'grey water' recycling.

Display Energy Certificate
How efficiently is this building being used?

HM Government

Arundel Building
Sheffield Hallam University
122 Charles Street
Sheffield
S1 2NE

Certificate Reference Number:
0551-0011-0280-4990-4092

This certificate indicates how much energy is being used to operate this building. The operational rating is based on meter readings of all the energy actually used in the building. It is compared to a benchmark that represents performance indicative of all buildings of this type. There is more advice on how to interpret this information on the Government's website www.communities.gov.uk/epbd.

Energy Performance Operational Rating

This tells you how efficiently energy has been used in the building. The numbers do not represent actual units of energy consumed; they represent comparative energy efficiency. 100 would be typical for this kind of building.

More energy efficient

A 0-25
B 26-50 ◀ 46
C 51-75
D 76-100
100 would be typical
E 101-125
F 126-150
G Over 150

Less energy efficient

Total CO_2 Emissions

This tells you how much carbon dioxide the building emits. It shows tonnes per year of CO_2.

Electricity
Heating
10-2008 10-2009

Previous Operational Ratings

This tells you how efficiently energy has been used in this building over the last three accounting periods

10-2009 46
10-2008 44
0 50 100 150 200

Technical information

This tells you technical information about how energy is used in this building. Consumption data based on actual meter readings.

Main heating fuel: District Heating
Building Environment: Heating and Mechanical Ventilation
Total useful floor area (m^2): 3150
Asset Rating: Not available

	Heating	Electrical
Annual Energy Use (kWh/m²/year)	74	64
Typical Energy Use (kWh/m²/year)	254	80
Energy from renewables	0%	0%

Administrative information

This is a Display Energy Certificate as defined in SI 2007/991 as amended.

Assessment Software: ORCalc V2-00-02
Property Reference: 548240900001
Assessor Name: Mr Steve Lloyd
Assessor Number: NHER006538
Accreditation Scheme: NHER
Employer/Trading Name: DomCom Energy Services Ltd
Employer/Trading Address: Hill Top House, 302 Ringinglow Road, Sheffield S11 7PX
Issue Date: 29-10-2009
Nominated Date: 01-10-2009
Valid Until: 30-09-2010
Related Party Disclosure: Not Applicable

Recommendations for improving the energy efficiency of the building are contained in the accompanying Advisory Report.

Figure 8.1 EPC for the Arundel Building, Sheffield

Intermediate/Can Be Measured Aspects of Sustainability

- Productivity (which we discussed in Chapter 6) can be measured in terms of sustainability features; for example number of days absent by a workforce can be compared in respect of a traditional air-conditioned building and one using more sustainable fresh cooled air (see Case Study 5).
- Carbon emissions can be problematic when making comparisons as alternative energy sources may have complex procurement routes which make true analysis of the carbon footprint difficult to measure accurately.

'Soft'/Intangible/Difficult to Measure Aspects of Sustainability

- Impact on corporate reputation, in an increasingly aware society a company's market share could be affected by its sustainability policies, especially if its target market comprises of Generation Y customers.
- Capacity to attract ethical funding, increasing popularity of funds targeting ethical and/or sustainable companies may provide access to competitive funding.
- Savings associated with retention of valued staff or attraction of more progressive staff due to an organization's proven sustainability record, which again may increase with significance as Generation Y employees dominate the workplace.

Beard & Roper (2006) sum up the potentials and problems of measurement with reference to a study by the California Sustainable Building Task Force in 2003 titled *The costs and financial benefits of green buildings*, which critically evaluated most of the literature available at the time examining the costs of building-related energy use, costs of emissions, water conservation, waste reduction, productivity and health and insurance/risk. The report attempts to definitively answer the question: is green building economically justifiable? The authors conclude that an initial investment of around 2% may yield a life cycle savings of ten times that investment. However, there is a small caveat about the financial benefit conclusions – energy, water and waste/emissions reduction savings can be calculated without a large margin of error, but health and productivity benefits are subject to a much higher degree of uncertainty. Further data and analysis is needed to quantify the true benefits/costs of sustainable building on worker well-being, employee turnover and health effects (Kats, 2003, p. v).

The challenge facing the corporate real estate manager and occupier is to develop the capability to measure and promote the value created by the more intangible aspects of sustainability. For example, encouraging more flexible working to reduce travel to work costs and space required within the office through the adoption, where appropriate, of hot desking and hotelling. Measuring the productivity improvements and cost reduction but also capturing the associated costs of working from home in the total environmental impact assessment. Most challenging to quantify are the health and lifestyle related benefits and how they may drive productivity, corporate loyalty and work-life balance.

Putting Sustainability into Practice

In the UK the Investment Property Forum (IPF) formed a Sustainability Interest Group (SIG) in 2006, to focus on the relationship between sustainability and the property investment market. The objectives of the SIG are to:

- inform and educate the property investment community on the impact of social and environmental issues and related policy responses on the value of property investments;
- identify and promote examples of best practice in responsible property investment;
- promote common standards pertaining to the definition of sustainable property investments;
- contribute to the development of relevant fund benchmarking measures.

Whilst their policy is largely directed towards investors it provides a clear and well articulated set of principles and deliverables that both corporate occupiers and investors can engage with.

- Energy
  - measure electricity, oil and gas consumption in Kw/h: CO_2 emissions;
  - target annual reductions in energy consumption: landlord and tenant;
  - target annual increases in renewable energy sourcing;
  - seek to ensure 'best of class' Energy Performance Certification;
  - benchmark energy performance for specific types of buildings and occupiers;
  - process of identifying poor energy performers and targets for improvement;
  - develop and implement procedures for working with occupiers to target carbon emissions reductions;
  - consider renewable energy options.
- Waste (incl. recycling)
  - measurement of performance data for waste and recycling across the portfolio;
  - develop and implement formal 'Waste Management' policies;
  - develop initiatives to assist tenants in waste reduction and improved recycling;
  - develop and implement building specific waste management plans.
- Transport
  - operating specific 'green'/sustainable travel plans;
  - policy of investment in public transport and reductions in car dependency;
  - active promotion of transport alternatives to tenants and employees.

- General management
  - support environmental standards and action plans for managing agents (ISO 14001);
  - develop sustainability guide for property acquisitions, lettings and disposals;
  - staff awareness and training programmes;
  - environmental procedures manual;
  - standard sustainability briefs for refurbishment design and construction including use of materials from sustainable sources;
  - ensure sustainability is incorporated within planned programme of maintenance;
  - annual reports that are fully integrated with company reporting process;
  - policy of public reporting of historic performance and disclosed targets;
  - ensure investment manager delivers best practice in its own occupied premises.
- Supply chain
  - ensuring suppliers have, or are committed to achieving, ISO 14001.
- Biodiversity
  - action plans to enhance biodiversity.
- Water
  - collate water consumption data to inform management plans, with annual reduction targets;
  - policies on grey water recycling, rainwater harvesting and sustainable urban drainage where appropriate.

CSR and Sustainability in Action

Marks & Spencer Plan A

A good example of a comprehensive strategy which blends CSR with sustainability and has a significant focus on corporate real estate is Marks & Spencer (M&S) Plan A. Plan A goes far beyond the rhetoric of many CSR statements and provides ambitious targets, aspirational visions and a detailed implementation plan, it reaches forward to consumers and backwards to suppliers.

This plan, launched in 2008, represents at the time of writing probably one of the most comprehensive strategies with a detailed set of 100 components which covers a wide ranging set of sustainable principles and stakeholder perspectives. The detail of the Plan can be found at the Plan A website: http://plana.marksandspencer.com. It includes green transport, reducing food air miles, reducing plastic bag usage, increasing organic suppliers and reducing the use of pesticides and sustainable procurement of food and other goods. We have selected from each section of the Plan examples of components that have a direct impact on or relevance to corporate real estate.

Climate Change

M&S aims to make all UK & Irish operations carbon neutral by 2012.

1. **Carbon neutral:** Aiming to make all UK and Republic of Ireland operations (stores, offices, warehouses, business travel and logistics) carbon neutral.
2. **Energy efficiency** (stores): Reducing the amount of energy in stores by 25% per square foot of floor space.
3. **Energy efficiency** (warehouses and offices): Achieving a 20% improvement in fuel efficiency and energy use in warehouses and offices.
6. **BREEAM:** Targeting all new stores to achieve (BREEAM) 'Excellent' rating and all other stores to be assessed against BREEAM rating system.
7. **Green electricity:** Sourcing or generating 100% 'green' (renewable) electricity for M&S stores, offices and distribution centres in the UK and Republic of Ireland.
9. **Green stores:** Having opened three 'green' concept stores in 2007, as well as incorporating many of the innovations into standard specification, plans to trial five Energy Stores and open three Sustainability Learning Stores.
10. **On-site renewables:** Having 20% on-site energy generation from renewables in all new builds where practicable.
22. **Green factories:** Having supported the development and opening of four 'green' factories with an aim to continue helping suppliers to open more 'green' production facilities.
33. **Construction waste:** Sending no waste to landfill from M&S store construction programmes.
34. **Construction and fit-out materials:** Working with WRAP (Waste and Resources Action Programme) to increase the amount of recycled materials used in the construction and 'fit-out' of stores.

Sustainable Raw Materials

This involves examining all the materials used in every area of the business from fresh food and fabric to the way shops are built. Aim to use materials from only the most sustainable sources protecting the environment and the world's natural resources for future generations.

65. **Water efficiency** (stores and offices): Reducing store and office mains water usage by 20%.
67. **Stores** (raw materials): Having developed sustainable raw material standards for store construction and equipment and aim to continue improving performance.

(**Source:** *Marks & Spencer (2008).* Plan A. *Retrieved from http://plana.marksandspencer.com accessed 05.09.09)*

Green Leases

Much of the work on green leases in the UK has been researched by Sustainable Environmental Improvements in the Commercial Sector (SEnvICS). SEnvICS is a European Regional Development Funded (ERDF Objective 2) project, part of Newport, UK's major regeneration programme. It aims to disseminate and encourage the uptake of a Good Practice Guide developed from two previous Carbon Trust funded projects, which were supported and assisted by King Sturge, Eversheds, RICS Foundation and the Environment Agency Wales.

The guide (*Incorporating environmental best practice into commercial lease agreements*) aims to help commercial landlords and tenants to incorporate a sustainable method, through the commercial lease agreement, to meet their environmental obligations, whilst improving the environmental performance and energy efficiency of the building in line with EU requirements.

Research has identified that the commercial lease agreement is often considered a systemic barrier to environmental improvement and resource efficiency within the commercial built environment, particularly in multi-tenanted buildings. Their research has identified good practice in a number of countries in the formulation and testing of green leases and provided guidance on their implementation.

The forward-thinking international legal practice Allen & Overy LLB in their guide to green real estate (2009) suggest the need for a hierarchy of green leases:

'Light green': suitable for existing dated or historic buildings to encourage operational and organizational improvements where physical improvements may be cost prohibitive or technically challenging.

'Mid green': suitable for refurbishments, with commitments for both parties and achievable and common goals to ensure a reduction in carbon emissions can be sustained. No compulsory penalties in place for not meeting targets.

'Dark green': suitable for new buildings with specific targets for reducing energy consumption, water use and waste disposal. These would be backed up with penalties on both parties if the targets were not met and rewards, agreed beforehand when met or exceeded.

Green Leases in the United Kingdom

Green leases are developing quickly in the United Kingdom. In 2007, a *Good Practice Guide* was published by CRiBE (Centre for Research in the Built Environment) in the Welsh School of Architecture at Cardiff University (Langley & Stevenson, 2007). The Guide gives example clauses that could be incorporated in a lease or schedule. The aim of this work is to encourage landlords and tenants to discuss the various clauses and pick those most appropriate for their circumstances. The following topics are set out below.

- New tenancy
- Renting premises
- Obtaining professional advice
- Financial matters
- Duration of lease
- Rent and VAT
- Service charge
- Repairs and service

- Assigning and sub-letting
- Alterations and changes of use
- On-going tenancy
- General communication
- Request for consents
- Release of landlord on sale of property
- Repairs
- Business rates
- Service charges

The SEnvICS project has supplied support and training to landlords and tenants interested in green leases. This has resulted in landlords investigating how they can proceed with green leases in a variety of multi-let buildings. Developments and case studies from this project can be found on the website http://www.greenleases-uk.com/senvics.php.

Hermes (a founder member of the Better Buildings Partnership), Exemplar Properties and Philips Research are the first companies in the UK to publicize the implementation of a green lease. Hermes has stated that it hopes to announce leases with other clients shortly.

Typical lease clauses in a Hermes and other green leases include the following:

- *sustainability and EPC – not to do or omit to do anything that adversely affects the EPC rating for the Property or the energy efficiency environmental performance or characteristics of the Property;*
- *so far as practicable to use materials from sustainable sources and to treat and maintain all materials fully in accordance with manufacturer's instructions and recommendations;*
- *to comply so far as practicable with the Owner's then current policy relating to responsible property investment;*
- *the Owner and the Occupier shall each co-operate with the other by the Occupier providing whatever information the Owner reasonably requires relating to the energy and water consumption and waste management statistics for the Property and for the Owner providing whatever information the Occupier reasonably requires relating to such statistics for the Common Parts in a reasonable manner in respect of any energy saving or carbon reduction initiative that the owner may choose to implement in relation to the Retail Park or that the occupier may choose to implement (but not so as to breach any of the Occupier's Covenants) in relation to its use of the Property;*
- *tenants have the right to install photovoltaic cells or panels or any other sustainable energy equipment ('the Equipment') on the roof of the Property.*

Green Leases in Australia

The Australian green lease is a self-contained generic document forming a legally enforceable management framework. The package includes five elements:

- agreed energy efficiency/Australian Building Greenhouse (ABGR) Rating;
- separate digital metering for each building/occupier;
- an energy management plan;
- a building management committee;
- dispute resolution.

The Australian Green Lease has been criticized for being dependent on building technology, and not flexible enough in its approach to building management. Accommodating separate digital metering could be problematic for older multi-tenanted buildings, which would need significant investment to rationalize electric wiring. It may also be an issue where the metered gas supply is used to provide unmetered heat and hot water to different occupants. Another issue is the requirement for a minimum ABGR rating – this is a responsibility of the landlord/owner, but can be difficult to achieve when the building is occupied by a 24/7 tenant.

Since there is little scope to vary the lease to meet specific requirements, eight versions have been written to encompass the major possibilities, as set out below:

Schedule A1 (gross lease – single occupant)
Schedule A2 (net lease – single occupant)
Schedule B1 (gross lease – dominant occupant)
Schedule B2 (net lease – dominant occupant)
Schedule C1 (gross lease – subsidiary occupant)
Schedule C2 (net lease – subsidiary occupant)
Schedule D1 (gross lease – voluntary)
Schedule D2 (net lease – voluntary)

The Australian government has been a major driver for the Australian Green Lease 1, although other clients include superfunds, telecoms, banks and property trusts. Detailed information has been made available on two of the buildings (60L and The Bond), which have been occupied under this scheme.

There are a variety of sources for information but the best is considered to be the Australian Government's Energy Efficiency in Government Operations (EEGO) website, from which copies of the leases' variations can be downloaded. This website also holds guidance notes on the green leases scheme.

Green Leases in Action in Australia: Investa

Investa is rated as the leading company on the Dow Jones Sustainability World Index (DJSI World) in the financial services super sector and operates businesses that include Australia's largest listed office portfolio. They state that:

> *we initiated the Green Lease Guide to help tenants who share our commitment to creating highly productive and environmentally friendly workspace better appreciate the range of benefits that can come from collaboration.*

Investa has produced a Green Lease Guide for commercial office tenants which works with their precedent lease, which is one of the most comprehensive green leases in Australia and provide inspirational material for anyone considering a green lease arrangement either as a tenant or a landlord.

The lease and accompanying guide are very comprehensive documents and we can only provide a very brief overview within this chapter; we do however recommend that you consult the original documents at www.investa.com.au.

Set out below is a small part of the Investa Green Lease Guide which deals with Investa's commitment to its tenants and which is reinforced by covenants within their green leases:

- comfortable, productive and healthy indoor environment;
- HVAC regularly tested for contaminants, and contaminants removed;
- regular indoor air quality testing;
- low energy use and greenhouse gas emissions;
- energy management guarantee for tenant (caps on bills and greenhouse gas emissions);
- monitoring and reporting of tenancy energy use and greenhouse gas emissions;
- sustainable and healthy transport options;
- proximity to a range of public transport options (high number and frequency of services);
- secure bike storage;
- showers, change facilities and lockers for cyclists;
- low potable water use;
- sub-metering of major base building water uses (e.g. cooling towers, bathrooms etc.);
- regular monitoring and reporting of base building water use;
- recycling of office waste;
- facilities for separate storage and recycling of paper, cardboard, containers and food waste;
- regular monitoring and reporting of waste going to landfill;
- sustainable cleaning services;
- achieves an environmental benefit without costing more;
- maintains good indoor environment quality;
- demonstrates environmental leadership and concern for employee well-being;
- green lease certificate issued annually to tenants;
- building user guide for tenants;
- environmental management plan for the building;
- regular reporting to tenants on base building environmental performance;
- formal mechanisms for gathering tenant feedback (e.g. regular surveys);
- dedicated contact for tenants within the building management staff.

The guide also provides comprehensive information on how the tenant can add value through the use of their offices in terms of fit out and design and selection of office equipment and furniture; and office management and operations to support sustainability. Investa also produce a certificate (Figure 8.2) showing which of the landlord and tenant commitments has been signed up to within one of their buildings, allowing occupiers to proudly demonstrate how they have worked in partnership with their landlord to apply green lease principles.

*(**Source:** Investa Property Group (2007).* Green Lease Guide. *Retrieved from Investa http://www.investa.com.au/Common/Pdf/Sustainability/GreenLeaseGuide.pdf accessed 5 September 2009)*

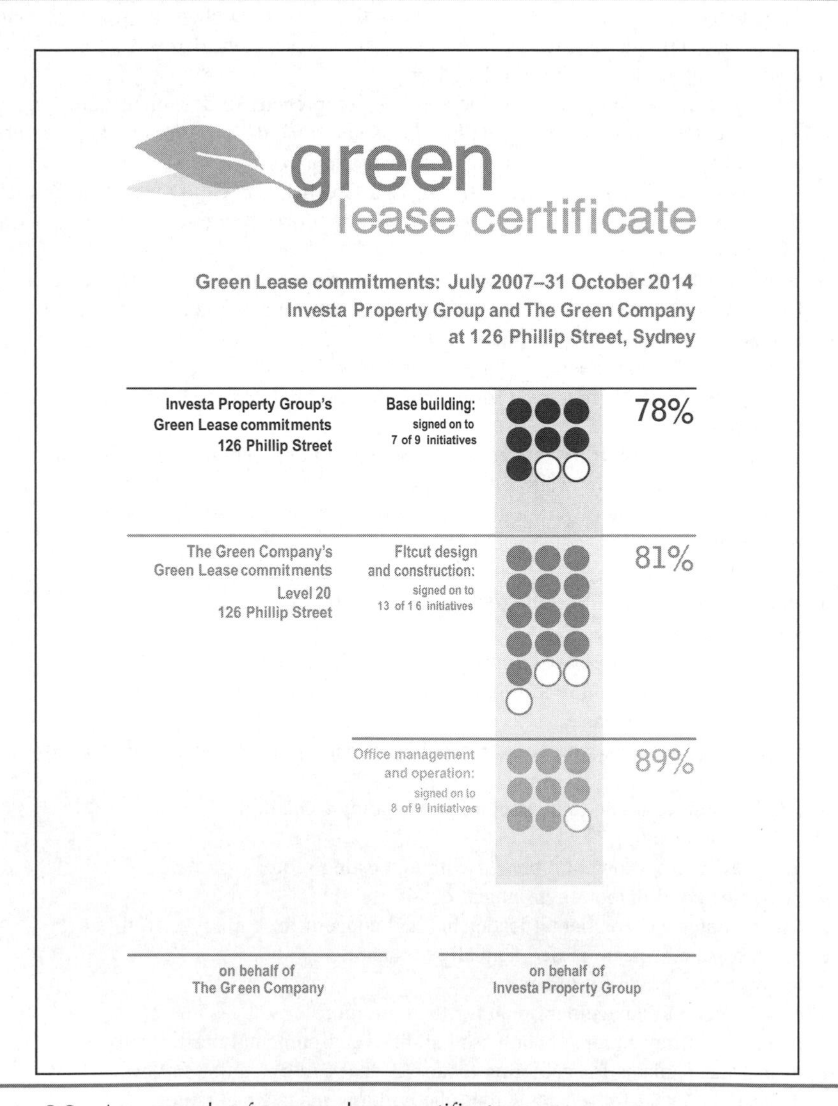

Figure 8.2 An example of a green lease certificate

The Carbon Reduction Commitment

Carbon reduction is a worldwide issue that is complex, political and subject to a variety of approaches and timescales. It originates from the Kyoto Protocol and countries worldwide are at different stages of implementation; the UK's scheme is due to commence in April 2010.

In the UK the government (as at 2009) has set ambitious targets as part of the Climate Change Act 2008 to reduce CO_2 emissions by 60% by 2050. The aim of the Carbon Reduction Commitment is to reduce the level of carbon emissions currently produced by the larger 'low energy-intensive' organizations by approximately 1.2 million tonnes of CO_2 per year by 2020. Buildings account for a very significant component of greenhouse gas production and the UK Department for Communities and Local Government (2008) estimates that 50% of greenhouse emissions in the UK originate from commercial buildings, and the Carbon Trust has estimated that 75% of commercial buildings that exist today will remain operational in 2030. Combining these two facts demonstrates the significance of commercial property and why the Carbon Reduction Commitment will be a key component to success.

In the UK the Department of Energy and Climate Change is responsible for carbon reduction and sets out the CRC Energy Efficiency Scheme (formerly known as the Carbon Reduction Commitment) as:

> *the UK's mandatory climate change and energy saving scheme, due to start in April 2010. It is central to the UK's strategy for improving energy efficiency and reducing carbon dioxide (CO_2) emissions, as set out in the Climate Change Act 2008. It has been designed to raise awareness in large organizations, especially at senior level, and encourage changes in behaviour and infrastructure. The scheme's amended title serves to better reflect the CRC's focus on increasing energy efficiency.*

The legally binding commitment will initially only affect larger businesses and public sector organizations not already regulated by a Climate Change Agreement or the EU Emissions Trading Scheme. It will apply, during the introductory period, to organizations that use significant amounts of energy defined as:

> *those organizations who are the counterparty to an energy supply contract (i.e. the organization is named on an energy bill) for supplies with at least one meter settled on the half hourly market (HHM) and its total electricity use through half hourly meters was greater than 6000 Megawatt hours (MWh) during calendar year 2008.*

At today's prices, this is roughly equivalent to total half hourly electricity bills of approximately £500,000 per year. The first compliance year April 2010 to March

2011 will be a reporting year only, with no sale of allowances. Organizations with HHM but supplies of less than 6000 MWh will not qualify, but will have to disclose information during the registration period. It is estimated that on introduction of the scheme about 5000 organizations will become full participants and 20,000 requiring information disclosure.

The scheme is organized in seven-year phases and whilst this may seem a long time for those organizations not falling into the introductory or second phases, companies should be aware of the significant impacts CRC will have on the sale and leasing of property and should be preparing for its impact, which will be significant, especially if the organization is occupying inefficient, high emission real estate. Major landlords, where their energy use is aggregated, may become part of the introductory phase and therefore tenants should be prepared for some CRC costs being recharged in service charge agreements. Bruntwood, the subject of our Case Study 1, will be within the first phase and is already well advanced with its preparation activities.

Organizations that occupy premises on green leases should benefit by having a pro-active partnership committed to reducing energy costs and omissions which should benefit them in terms of future CO_2 liabilities.

In simple terms the scheme involves the purchase of allowances – initially at a fixed cost, referred to as a 'safety valve' price set at £12/tonne CO_2 – from the government to cover anticipated CO_2 emissions calculated through estimated energy usage. In future years a 'market' will be created with allowances being traded at auction. The numbers of allowances and credits available will be restricted over time, thereby increasing the price of CO_2 and ensuring that landlords and occupiers continually examine energy use and work towards more efficient solutions.

Whilst the detailed workings of this legislation are outside the scope of this text, we recommend that ***all*** organizations start their preparations, even if they are not in the first tranche of organizations that the provisions apply to.

Conclusions

It is clear from the above that sustainability and CSR are extremely important issues for CREAM and that legislation, compliance and commitment will continue to increase.

The latest research at the time of writing, by Eicholtz, Kok & Quigley (2009), sponsored by the RICS, has examined the motivation for occupying green space in the United States and identifies four reasons why companies should occupy green buildings. The results validate our own views and reinforce what is presented in this chapter. Set out in Table 8.4, they form a very suitable conclusion to this chapter.

TABLE 8.4 Reasons Why Companies Should Occupy Green Buildings

Direct economic benefits:	Occupying a 'green' building can have direct economic benefits for a company through a number of means. First, and most straightforward, firms may simply be able to save money through lower occupier costs. This, however, depends on the nature of the lease agreement that the occupier has with the landlord. Another benefit can be through increases in employee productivity, and several studies have shown a correlation between a building's internal environments (e.g. its air quality) and employee health and productivity (This is examined and confirmed in our Case Study 5)
Indirect economic benefits:	More difficult to measure, but still of importance, are indirect benefits that firms may receive from leasing 'green'. One benefit relates to reputation, an increasingly important issue for organizations. A decision to occupy a 'green' building can help send a message to the outside world about the responsibility and commitment of the organization. There are good business reasons for doing this – if leasing 'green' office space leads to a superior reputation, it helps firms to attract investors more easily and at better rates. Secondly, firms don't just need to think about their investors – they also need to think about their customers, who are increasingly judging companies by their actions as well as their products. So, a superior reputation can help attract and retain customers. This can also be done to help offset or mask a controversial reputation, such as that of organizations in the tobacco or weapons businesses. Third, and related to the point about employee productivity, locating operations in a 'green' building may help a firm to attract and retain a loyal workforce
Risk avoidance:	Globally, the trend is towards increased environmental regulation, with stricter controls on a wide range of factors, such as waste emissions and energy use The likelihood is that this trend is going to continue and intensify so, in order to minimize the risk of running foul of ever-tighter regulations, organizations see benefit in occupying buildings that are likely to conform to environmental regulations for some time to come
Ethical behaviour:	There are some organizations, particularly in the government and not-for-profit sectors, likely to want to occupy 'green' space because it is the 'right thing to do'. Of course, some of these organizations may face softer budget constraints and may be in a position to pay a premium

Summary Checklist

1. Occupy a certified building such as BREAM Excellent or Very Good or LEED Platinum or Gold. Whilst anecdotal evidence is demonstrating that higher rents and more value may be attributable to these there is very limited evidence yet that these buildings will actually deliver the savings as promised by the measurement tools. For corporate occupiers a study of the TOTAL occupancy costs (as demonstrated in the Amazon Court building in Case Study 5) will only reveal whether or not real savings will be made.

2. Occupy a building with a good EPC rating – A or B. Again the same caveats may apply as above, although it is easier to measure the benefits in energy saving between different classes of rated buildings. The problems lie in the accuracy and credibility of the measurement systems. However, if energy costs do rise in line with expectations consideration of energy consumption is recommended as one of the most practical ways to save cost and be more sustainable.
3. Occupy a building with a responsible landlord under a green lease.
4. Introduce or retrofit sustainable measures – these can include solar shielding in hot countries, following the Adobe model of using trees to screen buildings or through reflective glazing or blinds. Other examples would be rainwater collection to use for grey water which is not only sustainable but, where water is metered, will reduce costs.
5. Manage the resources more effectively, install motion detectors to automatically switch off lights in unused areas, replace halogen bulbs with LEDs; provide cycle racks and shower facilities to encourage the use of bicycles.
6. Check for consistencies between your CSR statements and your sustainability position – can you celebrate sustainable solutions and policies as Siemens have done in their annual CSR report or do you have jarring inconsistencies such as your CSR statement next to your poorly rated EPC certificate?
7. Do you need the space you occupy? Undertake space and utilization studies (see Chapter 5) and downsize your occupation to better meet the measured needs of your organization thereby reducing wastage incurred in heating/lighting/cooling underutilized space.
8. Use ISO14001 to review compliance with legislation and build an environmental management system for your organization.
9. Take the free Environmental Health Check provided by the IPD at www.ipd.com/sustainability.
10. Consider your position in respect of your countries Carbon Reduction Commitment. In the UK get hold of a copy of the guide for landlords and tenants produced by members of the Green Property Alliance (2008) downloadable from the British Property Federations website: www.bpf.org.uk/topics/document/23672/carbon-reduction-commitment-crc-a-guide-for-landlords-and-tenants. If you are covered by the UK commitment you need to act to establish your liability. How you may reduce it in the future and how the commitment will affect you as a landowner, landlord and/or tenant.

References

Allen, & Overy, L. L. P. (2009). *A guide to green real estate*. http://www.allenovery.com/AOWeb/binaries/53237.pdf. Accessed 20.11.09.

Beard, J., & Roper, K. (2006). Justifying sustainable buildings – championing green operations. *Journal of Corporate Real Estate, 8 (2)*, 91–103.

Brown, R., & Wolf, E. (2008). Reclaiming the future: the culture must make some big readjustments. *The Next Agenda, Autumn 1988*(8).

Bruntland Commission (World Commission on Environment and Development. (1987). *Our common future.* Oxford: Oxford University Press.

Eicholtz, P., Kok., & Quigley, J. (2009). RICS Research: Why do companies rent green? Retrieved from: http://www.rics.org/site/scripts/download_info.aspx?fileID=5071&categoryID=523. Accessed 20.11.09.

Gensler. (2008). Faulty Towers: Is the British Office Sustainable? Report retrieved from: www.gensler.com. Accessed 05.09.09.

Green Property Alliance. (2008). *The Carbon Reduction Commitment: A guide for landlords and tenants.* Retrieved from British Property Foundation: http://www.bpf.org.uk/topics/document/23672/carbon-reduction-commitment-crc–a-guide-for-landlords-and-tenants. Accessed 05.09.09.

Hawken, P. (1994). *The ecology of commerce.* New York, NY: HarperCollins.

International Energy Agency. (2005). Electricity Information 2005. OECD.

ISO. (2009). *ISO 14000 essentials.* Retrieved from: http://www.iso.org/iso/iso_catalogue/management_standards/iso_9000_iso_14000/iso_14000_essentials.htm. Accessed 05.09.09.

Kats, G. (2003). *The costs and financial benefits of green buildings: A report to California's Sustainable Building Task Force.* Washington, DC: Capital E.

Langley, A., & Stevenson, V. (2007). *Incorporating environmental best practice into commercial tenant lease agreements: Good practice guide - Part 1.* Cardiff: Welsh School of Architecture.

LEED Retrieved from United States Green Building Council: http://www.usgbc.org/ Accessed 05.09.09.

McIntosh, A. (2009). *European property sustainability matters.* London: King Sturge.

RICS. (2007). *Surveying sustainability: A short guide for the property professional.* London: RICS.

Savitz, A. (2006). *The triple bottom line: How today's best-run companies are achieving economic, social, and environmental success – and how you can too.* San Francisco, CA: Jossey–Bass/Wiley.

Senge, P. (2008). *The Necessary Revolution: How Individuals and Organizations are Working Together to Create a Sustainable World.* New York, NY: Doubleday.

US Environmental Protection Agency. (2009). Retrieved from: <http://www.epa.gov/greenbuilding/pubs/about.htm> Accessed 12.11.09.

Further Reading

Cadman, D. (2000). *The vicious circle of blame. Cited in: M. Keeping (2000) What about demand? Do investors want 'sustainable buildings'? published by The RICS Research Foundation.* Retrieved from RICS: http://www.rics.org/Practiceareas/Builtenvironment/Sustainableconstruction/what_about_the_demand_20000101.html. Accessed 05.09.09.

Hawken, P., Lovins, A., & Hunter, L (1997). *Natural capitalism: Creating the next industrial revolution.* Snowmass: Rocky Mountain Institute.

Jaeger, J. K. (2001). Double dividend reconsidered. *Association of Environmental and Resource Economists Newsletter.* November.

Jansen, M. (2008). *Hermes lays down green lease gauntlet.* Property Week, 4 April, (p. 70.)

Jansen, M. (2008). Degrees of indifference. *Property Week* Sustainability Supplement, 2 May, (pp. 4–5.)

Jayson, P. (2008). Green leases – barriers or green hurdles? A lawyer's perspective. Presentation at: *Don't Forget Your Greens.* Newport, UK, 4 March Powerpoint downloadable at. http://www.greenleases-uk.com.

Kheel, T. (1992). *The Earth Pledge.* New York, NY: Earth Pledge Foundation.

Lust, G. (2007). Green leases. *Insight* (Property Week Supplement), 23 November (p. 63.)

Morton, S. (2003). Business case for green design, Building Operation Management, November.

Pinsent Masons. (2007). *Sustainability Toolkit Series: Part 3 - The Development of Green Leases*. Retrieved from. http://www.pinsentmasons.com/media/1153939890.pdf. Accessed 05.09.09.

Roper, K. (2003). *LEED and environmental issues in facility management*. Georgia Institute of Technology, 22 October.

Stallworth, H. (1997). *The Economics of Sustainability*. Washington, DC: US Environment Protection Agency.

Symons, D., & Williams, S. (2007). *The Green Lease Guide*. Boodle Hatfield. London: WSP Environmental.

Case Study 1

Customer Focused Real Estate: The Bruntwood Way

About Bruntwood

Bruntwood is a privately owned commercial property company established in 1976 by Mike Oglesby and a business partner (who lived on Bruntwood Lane). The first acquisitions were factories in the Manchester region refurbished to provide serviced accommodation. Following the industrial recession of the mid-1980s Bruntwood shifted into the office property market. The first office building Bruntwood bought was South Central in Manchester City Centre. Chris Oglesby joined Bruntwood in 1991 and was appointed Chief Executive at the end of 1999.

Today, Bruntwood owns over 90 buildings in Manchester, Leeds, Liverpool and Birmingham. Bruntwood has enjoyed unparalleled growth; starting with only £40,000 of private equity in 1978, its asset value topped £1 billion in recent years, before the global decline in values. It remains much lower geared and financially robust than most property companies, despite the challenging financial times. Bruntwood is an exceptional company, unlike any other property company, with an extraordinary commitment to customer service, which results in tenant retention of three times the national average. It does not use the word *tenant*; use of this word is reported to have resulted in a contribution to a swear box for many years and the refusal to use the *'t word'* in leases initially created some problems of acceptance with solicitors. To us, it reflects the Bruntwood commitment to treating customers differently. Bruntwood consistently ranks first in the Real Service Best Practice Index for Benchmarking Real Estate Excellence.

The Bruntwood business was initially built upon refurbishment of existing buildings; it has now recently completed new build offices, including One New York Street, Manchester (Figure C1.1).

This case study examines its customer focused approach and the benchmarking systems used to measure and constantly improve its offer. It is based upon information kindly provided by Bruntwood and an interview with Rob Yates the Customer Service Director and John McHugh, Customer Service Development Manager, in November 2009.

Corporate Real Estate Asset Management. ISBN: 978-0-7282-0573-4

Figure C1.1 Number One, New York Street, Manchester, UK

Bruntwood's Success

The exceptional success of Bruntwood, as we see it is, is set out below and connected to comments made by Rob Yates, during the interview.

- Buying well, spotting opportunities and recycling equity created through innovative and creative redevelopment to finance future projects:

  we are perceived as big risk takers but in fact if you look at us we very rarely take big risks.

Figure C1.2 A typical high quality reception area in a Bruntwood office building

- Recognizing that this is a cash business and creating a company around this so that customer satisfaction leads to retention of customers and therefore drives the cash flow:

  customer service, isn't just about happy smiley faces, and pulling in competitors' customers. It started with the initial approach that came from America, it's very much the flexibility of lease model, and that has been always the same, we've never changed it.

- Providing customer flexibility and a range of products from virtual offices, meeting rooms, services offices and large lettings:

  we have many examples of customers moving through our product range, from virtual offices to serviced offices to small office lettings to large office lettings; which Regus cannot do; our model is to have a range of services and products which feed into one another.

- Knowing what stock is required to grow your customers and being able to provide it at the right time:

  we understand the markets which we work in, we have the biggest sales team in the north west, people don't realize that, but it means we are geared up for making sure we understand our sales to stock requirements ... if you haven't got enough stock through the factory door, you're not going to be able to grow anymore.

A significant part of this success is attributable to the creation of a brand driven *by* and recognized *for* customer service. This customer service is driven through a number of sustaining factors including:

- recruitment of customer service staff from the hotel and hospitality industries where quality of service is central to the difference between the best and the rest;
- empowerment of staff, particularly general managers in each building; and
- a commitment to continuous measurement and benchmarking of customer service and acting upon the results.

It is interesting to note how Rob Yates confirms this approach:

> ***we have more people in our buildings than we do in our offices,*** *many of these people are trained, in the hotel industry, they are used to forming relationships with customers and that relationship is the biggest single factor of customer retention.*

Bruntwood publications set out the customer service team approach:

> *Our Customer Service teams make sure our buildings run to the highest standards both behind the scenes and front of house. For these people, delivering outstanding customer service comes naturally. There are dozens of anecdotes about the lengths our service teams will go to for the benefit of our customers. From dressing as waiters when one was let down by a catering company, to helping with jump leads when a car failed to start, to simply changing light bulbs and ordering taxis. The list goes on but the commonality remains – they'll do all they can to help.*

In the interview Rob Yates illustrated the contrast between their approach and most traditional property companies:

> *At development (X) the management is outsourced, compared to us, we insource our people. With a traditional landlord, how often do you see the property manager? If he/she is acting on 200 properties in the North West, you might see them once in two years; and who employs the people? There will be outsourced suppliers. Running a property [on this basis] you have outsourced the relationship and the people working day-to-day in the business.*
>
> *You would not find someone like Virgin or another hospitality industry or a hotel, who we try to model ourselves on, outsourcing their customer facing people or the people who form a relationship with that customer. Yet that is what the property industry does every single time.*

Bruntwood is not primarily driven by costs, as with many traditional property companies, but by cash. As Rob Yates states:

> *If we outsourced everything – Asset Management, Facilities, Finance etc – we could run this business with 10 people; and if you run this cost based model it makes a lot more profit. Bruntwood has 360 people, for a property company, it's a lot of people. We spend as much time looking after people, providing service, as we do the business.*

As a private company, Bruntwood has very different stakeholder expectations than, say, a REIT based company such as Land Securities. This focus on cash, the survival instinct of a family firm, has fostered the creation of a unique, customer focused property company and a brand that is rooted in openness, honesty, integrity, and a 'can do' customer centred approach.

The above also resonates well with a statement on the Bruntwood website:

> *Our customers have told us that the essential elements of our promise are a wide choice of offices, commercial flexibility, customer service and long term value for money.*

Whist this case study focuses on the benchmarking and customer recommendation survey it is worth noting some of the other features that we attribute to Bruntwood's success. These were discussed in the interview and provide in themselves a benchmark against which occupiers should measure their existing or proposed offer of service:

- being ahead of both the competition and regulation, for example Bruntwood's leases are not only compliant with, but actually exceed the standards of both the Code for Leasing Business Premises and the RICS Service Charge code;
- offering immense flexibility and the ability to 'tear up' a lease and start again;
- smoothing service charges to avoid financial shocks;
- employing general managers of buildings and other service customer facing staff trained in a service environment and motivated through performance systems linked to customer feedback and recommendation;
- future proofing both its own business and that of its customers by early planning for sustainability, including the Carbon Reduction Commitment Energy Efficiency scheme and the impacts it will have for both landlords and tenants;
- creating a Bruntwood green lease to deliver competitive advantage through a sustainable approach;
- recognizing the importance of energy security, including the generation of electricity for its own buildings.

Bruntwood continues to innovate, with initiatives like Red Rooms, 'pay as you go' meeting rooms that can not only be used by Bruntwood customers but other businesses. As with all Bruntwood products, the focus is on cash, raising revenue from difficult to let spaces on ground floors of buildings, offering an added value service to existing customers as well as a marketing opportunity for new customers.

> *We recognized a market place just for meeting rooms and re-branded a sub brand of our serviced space as Red Rooms. It acts as a focus for new external customers who are not in our serviced space or our buildings. An external customer who comes to a meeting for the first time sees Bruntwood around them, experiences the service quality and, who knows, in five years' time, when starting a business of their own may just register a bell and come to us as a customer. It is another distribution channel for sales leads.*

Bruntwood's Values

It is clear from every interaction with Bruntwood; through its documentation, website, or with its staff, that it has a very different approach, and this is evident when examining its values.

- **Our customers.** Although we own over 90 office buildings it's our 1000 customers who make our world spin round. Our customers are our lifeblood and we will do all we can to keep them happy.
- **Our people.** Our people make our business prosper and we want to invest and develop in them as well as our buildings. We are an accredited 'Investor in People' – this is the National Standard which sets a level of good practice for training and development of people to achieve business goals.
- **Our community.** Bruntwood's success is linked with the success of the regions in which it operates. We help our communities thrive by giving 10% of our annual profits to charity, encouraging our people to volunteer their time and skills and fundraising for local charities.
- **Our environment.** We all have a responsibility to reduce our impact on the environment. At Bruntwood we take this very seriously. Initiatives range from encouraging our customers to use our recycling scheme, partnering with the Eco Cities project, installing smart meters in our buildings to help reduce wasteful energy use and using environmentally sound suppliers.

Bruntwood's Commitment to Benchmarking Customer Service

We believe a significant part of Bruntwood's success is attributed to its commitment to high level benchmarking of its activities; a genuine attempt to integrate the benchmarking process into its strategy; and to act upon and value all feedback received. Bruntwood does not value surveys such as the Occupiers Satisfaction Index. It looks at best practice from other industries and utilizes both strategic tools such as the Balanced Scorecard (as discussed in Chapter 5) and recommendation surveys used for many years by companies such as Enterprise Cars and used in luxury service driven organizations such as Lexus.

We examine, in this case study, two separate benchmarking activities that Bruntwood engages with.

The REAL SERVICE Best Practice Benchmarking Service

REAL SERVICE is a member organization that provides a forum for sharing and benchmarking best practice in customer service within the property industry. It measures how far each member organization has progressed in the adoption of best practice in customer focused property management. The REAL SERVICE Best Practice Index has 12 members who provide evidence against 25 criteria organized under six building blocks.

It is interesting to compare the building blocks with those developed by BAA, discussed in Chapter 4.

- **Building blocks of the REAL SERVICE best practice benchmark:**
  - Service Strategy
  - The Occupier Customer
  - Products and Services
  - People and Leadership
  - Business Operations
  - Measurement

The annual REAL SERVICE report provides a report on each building block containing the following:

- **Performance Status.** This shows how the organization has performed and how this score has changed over the last year.
- **Key Findings.** This summarizes the progress the organization has made to improve best practice based on evidence provided and reviewed by REAL SERVICE.
- **Member Actions.** This is a record of some of the actions that the organization plans to take in the next 12 months to enhance best practice.
- **Recommendations.** Drawing on the best practices of other REAL SERVICE members who may wish to be considered.

Bruntwood has been ranked number one consistently within this practice group and its evidence provides outstanding achievement across the majority of the categories. The benchmarking process does provide annual recommendations for the company, for example:

> *Utilize your expertise to help more occupiers achieve their own environmental targets*

and seeks to provide a process for continuous monitoring and improvement.

However, as Rob Yates comments:

> *We got nearly 94% in the last Real Service benchmarking exercise. Apart from a couple of technical points with regard to the lease code of conduct there was nowhere for us to go, so we are not really getting a lot back. A lot of the other companies in the index cannot adopt our business model.*

Whilst Bruntwood is clearly in advance of the competition in respect of customer service it does utilize all the feedback from the index and visit other companies to compare its processes against. For example, British Land and their facilities management operations at Broadgate are recognized as excellent and Bruntwood visited them to learn from the way in which they do things.

The Customer Recommendation Survey

The following information is extracted from material kindly provided and written by John McHugh, together with a copy of the 2008 Bruntwood Customer Recommendation Survey.

The customer recommendation score concept was developed by loyalty business model expert Fred Reichheld of Bain & Company and is discussed in his book *The ultimate question: Driving good profits and true growth*, based on the link between customer satisfaction, customer loyalty and profitability.

A customer's response to the 'recommend' question serves as a strong indicator of customer satisfaction. According to Reichheld's research, customers with higher scores typically buy more products, remain customers for longer and recommend the service to others.

The Bruntwood customer recommendation survey asks every customer's key decision maker only one so-called ultimate question:

> *Please rate the likelihood you would recommend Bruntwood to another business, friend or colleague.*

Responses to the customer recommendation survey are recorded on a 0–10 scale, with 0 meaning the least likely to recommend and 10 meaning the most likely to recommend. This approach is very different, and more brutal than traditional surveys, such as the Occupiers Satisfaction Index (OSI). Only scores of 9 or 10 equate to a recommendation and in Bruntwood's eyes only these scores guarantee retention of that customer. A score of below 7 is a detractor and 7 and 8 neutral.

The difference between the percentage of recommendation and detractors is the customer recommendation score. For example, if 50% of a building's customers respond with a 9 or 10, and 30% respond 0–6, the building's score would be 20%. The survey is managed externally to ensure objectivity and to allow customers the opportunity to provide feedback in confidence.

Bruntwood's net recommendation score in 2008 was a very impressive 26%. As a general rule of thumb UK companies are found to have a net recommendation score of around 5% with US companies now up to around the 13% mark. If the OSI was done on this basis we believe the results would probably demonstrate a net recommendation of around −3%. Mobile phone network companies have had recommendation scores as low as −18%, indicating more detractors than people who would recommend them. However, some car manufacturers such as BMW and Lexus have scores of over 30% and help to enhance a company's reputation and develop completely satisfied customers.

The recommendation survey plays an important part in Bruntwood's commitment to continuous improvement of customer service. The data can be examined in terms of overall performance, by region, building and categories. The categories allow greater detail to be examined within the context of the overall recommendation scores. Examples of categories examined, resulting in detailed customer comments and feedback are:

- FM: Structure and Fabric
- FM: Services
- Customer Service
- Lease Flexibility
- Lease Contract Issues
- Value for Money
- Product Quality
- Moving in Procedure
- Billing and Invoicing

Whilst Bruntwood has a very impressive recommendation score which underpins its retention rate of three times the UK average, it inevitably has some areas that score less than 7 and are considered a detractor. These are picked up in the survey and any individual comments examined and acted upon. As Rob Yates states:

> *We may still score low on some building issues – e.g. where a lift has not been performing well. But in fact we find that a few detractors can present an opportunity because we know we can pull the customer round, we know we can do it. We have had a number of customers who may have issues like*

air conditioning, in some of our older buildings, there have been problems, but they still score us with a 10 because of the way we are handling it.

Individual comments in the survey demonstrate very positive responses around the flexibility, value for money, first class, responsive service and welcoming, professional staff.

The survey, which is telephone based and targets key decision makers, has a very high response rate of 75%. It is followed up by a variety of formal customer interactions including focus groups, interviews, networking events, customer forums as well as the day-to-day informal interactions with the general managers.

Conclusions

The Bruntwood approach to real estate is unique in many ways. Its private company structure, responsible approach to risk and lending, ability to buy well, spot opportunities and undertake innovative redevelopment of office buildings has created solid growth and impressive financial performance. However, for us, it is the extraordinary commitment to customer focus, driven by the focus on cash and customer retention, that sets Bruntwood apart. The integration of robust benchmarking is one symbol of that commitment and an integral part of the strategic approach that defines Bruntwood as a very special real estate company and a landlord of choice.

We recommend tenants to compare their landlord and their customer focus against the Bruntwood model and we recommend asset managers to consider the strategic importance of customer satisfaction and reflect upon its impact on retention, cash flow and ultimately the value of real estate assets.

Case Study 2

Benchmarking: Actium Consult – A Fresh Approach to Real Estate

About Actium Consult

Actium Consult is an independent strategic management consultancy practice, which works with organizations to create accommodation strategies to deliver innovative property, workplace and people solutions. Actium Consult is a very interesting company for us as it is engaged in many of the strategic operations that we describe in this book. It is providing strategic management consultancy and CREAM through a unique combination of experienced professionals. It has a range of strategy consultants who provide business property strategy for owners and investors and accommodation strategy for occupiers. A lot of Actium's work is evidence based, using its own Total Office Cost Survey as a platform for many of its activities. We commend its strategic approach to benchmarking, which not only focuses on space and costs but also integrates many of our themes, including new ways of working, sustainability and productivity. This case study is based upon material from Actium Consults website and an interview with Andrew Procter, Managing Director and Bill Fennell Director of Research.

The Actium Mission

Actium states on its website:

> *We strive to be the leading Management Consultancy in property and accommodation. We understand organizational objectives and user requirements. We achieve individual solutions for clients which are:*

- Knowledge based
- Sustainable
- Practical and
- Independent

Corporate Real Estate Asset Management. ISBN: 978-0-7282-0573-4

Our scope is wide ranging and spans strategic financial advice including business case development and feasibility study work to a holistic accommodation strategy approach which encompasses in particular office accommodation strategy including managing organizational change, office change management and office occupancy cost benchmarking.

We deal with the hard issues such as estate strategy, property portfolio management, asset management, capital investment appraisal and office costs as well as soft issues such as change management and managing organizational change. Our in-depth experience in accommodation strategy includes work both for the private and public sector and we have delivered complex change management and relocation strategy projects from start to finish.

What we find interesting about this approach is the recognition of the importance of both hard and soft issues and a genuine willingness to integrate the two; whilst real estate and office benchmarking is at the heart of its approach, Actium brand themselves as management not real estate consultants. Its approach resonates with our strategic approach examined in Chapter 2. The way in which Actium practises also underpins our own teaching approach and validates our integration of real estate, facilities, business and change processes that makes our courses unique.

In the interview Andrew Procter indicated that they are very aware of the research around Generation Y and in their work in producing client briefs for relocations they feel that not enough attention is given to the needs of the next generation and that some companies spend lots of money on workstations that are inappropriate and potentially challenging to Generation Y graduates.

Providing Occupier Solutions

Essentially what Actium does is turn the occupier approach on its head compared to traditional real estate professionals by addressing the total picture of both property and people and by having expertise in managing organizational change. They recognize that the process of relocation or refurbishment of office space is as much about people as it is about property. Once a property solution has been found, which is where most real estate consultancy ends, for them the hard work starts in ensuring that the solution is effective, sustainable and aligned with the client's business.

Actium states on its website:

We have expertise in hard factors such as cost, location and layout and also soft factors such as culture, coaching and strategic change management. The practice has developed an impressive reputation for leading innovation, strategic management consultancy and research in the areas of performance measurement, outsourcing, property strategy, and rationalization strategy

and change management. The practice acts for property occupiers and service suppliers and has been engaged in work for an impressive client list in both the private and public sectors. Actium are property strategy consultants which help organizations, by spanning the gap in the corporate occupier market between theory and practice. Actium take an holistic approach.
Our occupier services include:

- Accommodation strategy
  - Alignment of business and property strategy
  - Accommodation audit and planning
  - Relocation and rationalization analysis
  - Programme management and delivery team procurement
  - Design management and briefing
  - Efficiency analysis for reproduction and printing requirements
  - Flexible working strategy
  - Space modelling
  - Utilization surveys
  - Accommodation protocols
  - Office cost analysis
  - New ways of working assessment
  - Making better use of office space

Actium and Change Management

Actium recognizes, as we do in Chapters 1, 2, 6 and 7, that introducing new technology, reconfiguring space and/or introducing new ways of working without appropriate protocols and change management processes can prove ineffective, create user frustration and actually diminish productivity leading to lost revenue. In essence it clearly recognizes the linkages between real estate, people and workplace psychology that we set out Figure 2.8, in Chapter 2. Actium states on its website:

The focus of much of our consultancy work is on how one can introduce new working practices into organizations so that it supports people in doing their jobs. We have considerable expertise in cognitive task analysis and a general analytical understanding of change management and managing organizational change so that we can advise organizations on exactly what the new resources must be capable of doing and on how change should be introduced into the workplace. We also have experience of how technology can best support distributed team working and also how organizational and team factors impact on effective performance, to advise organizations on effective strategic change management. Actium has expert behavioural and cognitive psychologists as part of their consultant teams to inform and guide effective change within client organizations. Our work in this area includes:

- Strategic change management
    - Training and staff development programmes
    - Motivational programmes
    - Evaluation of IT systems in terms of usability
    - Change and new ways of working programmes
    - Service delivery process mapping
    - Post occupancy staff surveys
    - Staff attitude surveys and awareness exercises
    - Staff satisfaction surveys
    - Home and mobile working initiatives
    - Work–life balance initiatives
    - Workplace behaviour protocols and working protocols
    - Happy office surveys
    - Document reduction programmes
    - Remote output based management

The Actium Total Office Cost Survey

Actium publishes an annual Total Office Cost Survey (TOCS) to assist occupiers, suppliers and property professionals with office cost benchmarking and cost reduction initiatives. The data underpinning TOCS is supplied by 18 leading data suppliers, including BT, Johnson Controls, AON Insurance and tpbennett.

We asked Andrew Proctor how it differs from the IPD approach:

IPD occupiers is basically peer group benchmarking for property, and is largely covert, users may know they are within an upper quartile but this may be impenetrable and difficult to extract true comparisons. What we have tried to do with TOCS is to allow users to look at the cost of workstations, and space in a particular location and begin to examine how their occupancy costs compare.

Some organizations say we can understand this [TOCS] because we can make comparisons with our portfolio and then use this as a base to start to make things happen.

IPD Occupiers can be very onerous. You have to answer a multitude of very detailed questions like; how long you have to wait in a queue for a photocopier; which has a tendency to make the results impenetrable.

We try to make ours simple, with clear definitions, transparent calculations and an ability to make real comparisons in a straightforward way.

The full survey, now in its 11th edition, includes:

- Full costs for brand new and also 20-year-old office buildings in 50 UK locations.
- Net Effective Rents and rent free periods being granted to new tenants in 50 locations (this data is the result of structured interviews with 120 agents throughout the UK).
- Full cost analysis (per m^2) of rent, rates, fit out, furniture, insurance, maintenance, moves, reinstatement, security, cleaning, utilities, energy, telephones, catering, reception, post, reprographics and management.
- A practical guide on how to use TOCS to measure the real total costs of offices.

The detailed breakdowns of costs for both a brand new office building and a 20-year-old building are based on stated standard assumptions including that the offices are:

- 5000 m^2 (NIA) self-contained, good quality and in the prime pitch for the location;
- of steel frame and curtain wall construction on four floors;
- held on a 10-year lease, fully repairing and insuring institutional lease;
- workstations are at an overall density of 10 m^2 per person with one workstation per staff member.

As with the IPD Occupiers benchmark discussed in Chapter 5, TOCS uses standard numbers of years to convert capital items such as furniture to an annual equivalent.

TOCS therefore provides a benchmark based on the standardized assumptions of both the total occupancy cost per square metre and per workstation; and a detailed breakdown providing costs per square metre NIA, per annum of:

- Net Effective Rent: Headline rent adjusted for incentives such as rent free using the Accounting Standards Board guidelines of straight line application up to the first rent review, e.g. for a 10-year lease with a review in year 5 and a rent free period of 12 months is equal to a 20% reduction per annum for 5 years up to rent review.
- Rates (UK Property Tax)
- Fitting Out (annualized over 7 years)
- Furniture (annualized over 5 years)
- Insurance
- Internal Repairs and Maintenance
- Mechanical and Electrical Repairs and Maintenance
- External Repairs and Maintenance
- Internal Moves

- Reinstatement
- Security
- Cleaning
- Waste Disposal
- Planting – Internal/External
- Water/Sewerage
- Energy
- Telephones
- Catering
- Reception
- Post/Courier
- Printing and Reprographics
- Management Fees

The TOCS document also provides a table for users to complete to compare an organization's costs against the benchmark and to create an action plan to manage costs that are in excess of the appropriate benchmark.

Actium Consult sets out how the TOCS Survey can be applied, and this is shown here in Figure C2.1.

We recognize this as an excellent foundation for application of strategic change and the approaches to CREAM adopted in this book. Benchmarking, in this case the TOCS results, can provide the catalyst for:

- performance measurement and benchmarking (Chapter 5);
- change and transformation of the workplace (Chapters 1 and 6);
- relocation (Chapter 7);
- changes to procurement and outsourcing (Chapters 3 and 4).

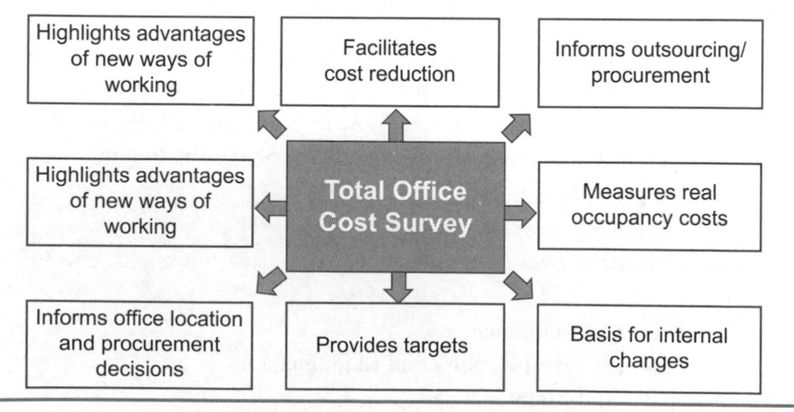

Figure C2.1 The use of the TOCS Survey as suggested by Actium Consult

Actium advised Land Securities on their Landflex buildings (see Chapter 3) within the London Portfolio to prepare a cost comparison to benchmark the Landflex product in respect of major buildings at Aldersgate, Eastbourne Terrace, Soho Square, Mark Lane and Empress State Building. These studies were used to validate the cost competitiveness of the Land Securities offering to their major customers.

TOCS is frequently used to create the business case which then underpins change and reconfiguration. So, for example, looking at an organization that has a number of inefficient or expensive buildings TOCS can be used to examine the business case for rationalizing to a single building. The costs and space savings can then be used to create a model to examine how new ways of working can be introduced to improve workflow, workspace satisfaction and ultimately productivity.

The Actium Consult Guide to Making Better Use of Office Space

Actium provides a number of useful resources on its website which demonstrate its practical, yet evidence based approach. Below we have replicated Actium's guide to making better use of office space:

As you are probably aware, one thing you can be sure of when running an office, is that things are always changing, even if you remain in the same premises.

Given the varying demands put on your office space, you need to make best use of every square inch. Then the day comes when you think that you have no alternative but to move to somewhere larger. Don't panic – you may not have to move. But you need to be methodical.

Find out how much space you need *– you will need to know some basic information such as:*

- How many people you need to accommodate at workstations
- Some idea of how much document storage you need
- A view on the amount of meeting space and training rooms required
- What catering provision you want
- What 'non-office' space, such as print rooms, lecture theatres and semi-industrial space, you need.

This doesn't actually help you use space more efficiently, but it does tell you how much space you theoretically should need, and if you have less than the model suggests, then being ultra-efficient in your use of space may not be enough.

Major space saver No. 1 *– reduce the number of cellular offices. Cellular offices are very inefficient use of space, and in many cases they are no longer necessary. The old days of needing a door between you and the pounding mechanical typewriters have long gone, it is possible to have discreet telephone conversations in open plan (or go to a meeting room to take/make a call); and now that so much more documentation is held on computer rather than in cupboards, there is much less confidential paper to lock up.*

Major space saver No. 2 *– get new, smaller desks. In the 80s and 90s there was a big increase in L-shaped desks – most office workers had computer terminals, and the corner of the L was needed to take the CRT monitor, but now, good sized flat screen monitors are easily available at affordable prices. As well as using less energy, the flat screen monitors allow a straight desk (with or without a comfort cut-out for the waist). This has 2 advantages for space utilization – firstly there is a saving of about half a square metre or 5 square feet in the footprint of a workstation. And secondly, the smaller workstations are less likely to suffer from 'fit-factor' issues, where the particular dimensions of a building mean you just can't get an extra desk in each row.*

Major space saver No. 3 *– document management and decluttering. This comprises:*

Having an off-site archive for old/seldom used documents
Getting rid of documents no longer needed and the cupboards they were stored in
Using modern high density storage (e.g. roller racking) for large quantities of documents required on-floor
Going electronic

Major space saver No. 4 *– have fewer desks than staff (desk sharing). This is the biggest space saver, but has been left until last because it is the most complicated. It depends on your IT and telephony systems being capable of supporting 'free-seating' or 'roaming' – i.e. users can 'log in' at any telephone or PC. The introduction of desk sharing also requires change management, and above all an improvement in the environment overall. However, many desk-sharing systems work on a ratio of around 8:10 (8 desks per 10 staff). The exact ratio will depend on the nature of the business, and the ratio may not be the same across the whole business – you might need a lower proportion of desks for your travelling sales team than for your accounts receivable clerks. However, a ratio of 8:10 results in 25% more staff accommodated at a given number of workstations, providing a big increase in capacity. Although it is not a cheap option, it can be a great deal less expensive and disruptive than an office move.*

Discussing the way in which Actium works with helping clients embed new ways of working and managing strategic relocation; again we see a lot of resonance with our approach. For example Andrew and Bill confirmed that:

> *some of the work we get involved in is with design briefing ... because architects are not very good at sitting down with clients and mapping business process and mapping what people are really wanting from their workplace, and what it will do for them. To do this the soft [people] work needs to be done as the same time as the hard [real estate] work.*

Actium has been involved in a project with Ealing Council who have reinvented its service delivery and instructed Actium Consult to programme-manage the accommodation work stream, which included:

- a 20,000 m^2 office HQ;
- a new ground floor 2000 m^2 walk-in centre;
- a new 150 seat contact centre;
- a training facility;
- a Registry paperless office solution.

> *In the Ealing project we worked with Architects Pringle Brandon and took a long time to really understand what the client wanted; local authorities are very complex animals and we worked hard to see how the business processes, workflow, the people and the workplace work together; this includes understanding Generation Y and the emotional support that Generation Y and other workers require. We aim to provide a detailed specification for architects to work with from the end user perspective, which not only is efficient, using TOCS, but also effective in promoting better workflow, communication and ultimately productivity.*

Again, the Actium approach resonates very closely to the strategic briefing process we outlined in Chapter 7 and with the need to understand generational change discussed in Chapter 1. Finally, the fact that Actium use a fully trained psychologist as part of their team to capture pre- and post-occupancy perceptions of the workplace underpins our examination of the changing workplace in Chapters 6 and validates what we set out in Figure 2.8, which incorporates psychology in our CREAM model.

Finally, we also recommend readers interested in examining new ways of working and how it may offer opportunities to improve the efficiency and effectiveness of your business, to request a copy of the Informative Guide in NWW Practices and Implementation Techniques from the Actium website at http://www.actiumconsult.co.uk.

Case Study 3

Workplace Transformation: Nokia Connecting People

About Nokia

Nokia is a pioneer in mobile telecommunications and the world's leading maker of mobile devices. Today, Nokia is connecting people in new and different ways – fusing advanced mobile technology with personalized services to enable people to stay close to what matters to them. Nokia also provides comprehensive digital map information through NAVTEQ; and equipment, solutions and services for communications networks through Nokia Siemens Networks.

Every day, well over one billion people, in more than 150 countries, use a Nokia. Today, Nokia is building on its leading position in mobile devices by bringing people personalized services and contextually relevant content to create a unique and compelling user experience. Nokia has mobile device manufacturing facilities in countries around the world and a strong research and development (R&D) operation that spans 16 countries. Nokia employs approximately 121,000 people globally.

Nokia's roots are in Finland, and their head office – Nokia House – is in Espoo, part of the Greater Helsinki area. Nokia House, as can be seen in Figure C3.1, consists of three buildings accommodating approximately 3000 employees.

In 2007, Nokia worked with Clive Wilkinson Architects and a local company, Gullsten and Inkinen, to envision a completely new environment in Nokia House, aligned with new working practices and expectations for future ways of working. The workplace transformation at Nokia House began with a three-stage refurbishment project.

In the fall of 2007 Nokia embarked on the first stage of the project. This included the refurbishment of space to provide a range of workspaces for mobile workers – employees who do not require a fixed desk – as well as areas for those who do. Nokia has undertaken a number of evaluations of the first phase of the refurbishment project, seeking feedback from both mobile workers and desk-based workers. In addition to surveys like these, Nokia has also conducted utilization studies to establish the amount of time that desks are occupied.

Corporate Real Estate Asset Management. ISBN: 978-0-7282-0573-4

Figure C3.1 Nokia House in Espoo, Finland

This case study is based on the lessons learned from the completed first stage refurbishment at Nokia House.

How Work Is Changing

The drivers of change in the workplace are not unique to Nokia and are valid for most global organizations. However, Nokia's positioning with regards to technology gives their employees a unique opportunity to respond to these drivers for change. Table C3.1 shows the shift from traditional ways of working to the new, emerging ways of working.

Work at the Office

Research undertaken by Nokia revealed that a typical desk in Nokia House is occupied less than 50% of the time. At any given moment, only 40% of employees are at their dedicated workstations. This means that Nokia employees are already engaged in mobile working, in one form or another. The space utilization statistics offer an interesting insight into space efficiency opportunities. If desks are occupied only 50% of the time and are dedicated to one individual, then the space 'asset' is not fully utilized. However, if desks are utilized as and when required then fewer desks would be required, hence increasing space efficiency. When not in the office, mobile

TABLE C3.1 From Traditional to Emerging Ways of Working

Traditional ways of working	Emerging ways of working
Work at the office	Working where and when needed
Performance measured according to time spent in the office	Performance measured according to results
Manager supervising	Manager mentoring and coaching
Team members in the same place	Virtual teams, mobile individuals
Space designed based on status and hierarchy and held *'just in case'*	Space designed based on functions and tasks and provided *'just in time'*

workers maintain their productivity levels by utilizing the latest tools and technology to work wherever they are.

Performance Measurement

As the workplace changes, so does the way that performance is evaluated. In today's competitive business environment it is not appropriate to measure employees' productivity based on how many hours they spend at a desk. The performance paradigm shift means that performance measures need to be based on outputs rather than 'face time' in the office. Productivity measurements cannot always be captured in a spreadsheet. For this reason Nokia chooses to include measures of perceived productivity, rather than seeking objective measures.

In addition to the issue of employee performance is the importance of integrating Nokia's values during workplace planning. The Nokia values are:

- engaging you
- achieving together
- passion for innovation
- very human

It is critical that these values are supported in all of Nokia's workplaces worldwide, and this is partly assessed through evaluating employee satisfaction and motivation/engagement.

Line Manager Role

An integral part of Nokia's shift to mobile working is the role of the line manager. The traditional role of line manager with an employee visibly accessible for 100% of their time (e.g. 40 hours per week) does not fit with the philosophy of mobile working. Instead, the typical mobile worker is:

- part of a number of different projects
- potentially a leader of a team/project
- supported by a mentor or coach
- a contributor to multiple knowledge communities

The complexity of the mobile worker's role means that the line manager needs to be more of a mentor and a coach rather than a task manager.

Team Members

The use of technologies such as mobile devices, Internet sites, social networking sites and virtual communities means that team members do not have to be in the same place at the same time in order to collaborate. This enables the mobility of individuals and allows virtual working. The use of virtual technologies is particularly useful when working with colleagues who are geographically dispersed.

Space Allocation

Giving Nokia employees the autonomy and flexibility to choose where and how they work means that workspaces need to be designed to support the variety of tasks that need to be performed. The mobile working concept leads to a *'just in time'* approach to workspace provision, since mobile workers can pick the most appropriate time and place to work. This represents a shift away from the traditional approach of space allocation based on status and hierarchy, where employees are allocated workplaces *'just in case'* they need it. Clearly, with only 40–50% space utilization on average, allocating individual space is a vastly inefficient use of space.

Target State for Nokia

The communications technologies sector is a very competitive and constantly evolving business environment. This means that for Nokia to maintain its leading position in the marketplace, it needs to constantly evaluate the way that its products are developed and its services are delivered. Nokia's goal in achieving this target state is to combine innovation and efficiency (Figure C3.2).

Nokia uses its own unique solutions to enhance business performance. Achieving the targeted state, illustrated in Figure C3.2, requires a more collaborative approach to problem solving, as well as a passion to deliver results through a new way of working. Collaborative tools are essential. The constant drive to be first to market is

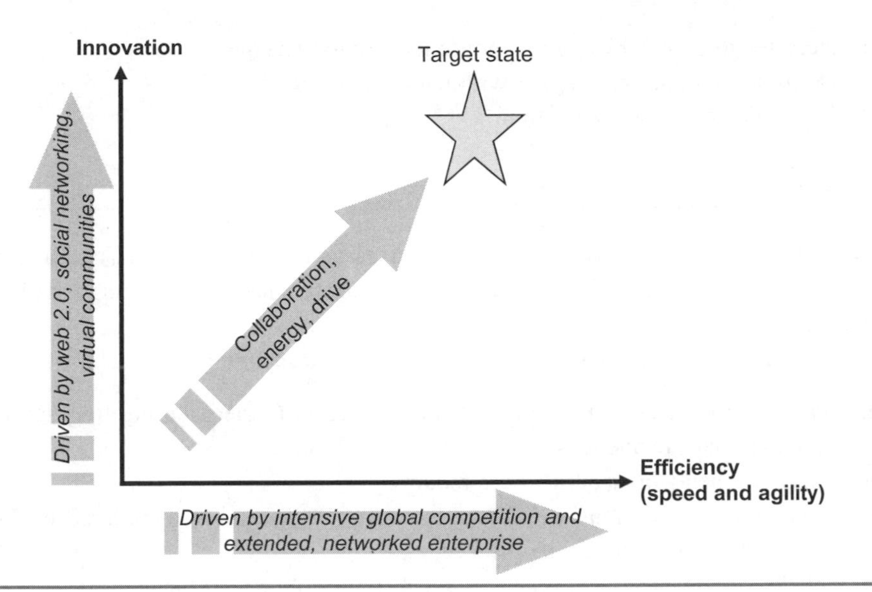

Figure C3.2 Target state for Nokia

supported by Nokia's commitment to both innovation and efficiency. It is this approach that provides Nokia with a competitive edge.

Mobile Workplace Concept

The emerging workplace trends allowing more collaborative and flexible ways of working, coupled with Nokia's new developments in communication technologies mean that Nokia is uniquely positioned to take advantage of mobile working. It is therefore appropriate to identify the drivers and enablers of the Nokia mobile workplace concept. In addition, we will explore the benefits of mobile working.

Drivers

Some of the drivers of mobile working can be classified as coming from the organizational perspective. Such organizational drivers include: a more agile response to the business environment, development of a unique brand and culture, and increased cost efficiency. Drivers for mobile working from the employee perspective include the need for more flexibility, and autonomy, in how and where work is done.

The drivers for the Nokia mobile workplace concept can be summarized as:

- Increased remote and mobile working – people are already working this way
- Employee demands for increased choice and autonomy for where and when they work

- Need for more flexibility in response to business change
- Desire to align physical space with culture and brand
- Cost efficiency/asset utilization.

Enablers

The physical space provided and the tools and technology available act as enablers to allow mobile working. The use of these resources needs to be supported by management practices.

The enablers for the Nokia mobile workplace concept are:

- shared workspaces plus increased other settings (variety of meeting spaces, informal areas, phone rooms, project team areas);
- some assigned workspaces where required;
- collaboration tools along with IT (video walls, PC-based video conferences, IM, social networking tools, telepresence, etc.);
- policies and management practices supporting mobile working.

Benefits

The benefits to Nokia of adopting the mobile workplace concept are as follows:

- space efficiency;
- accommodate more people over time without continuous reconfiguration;
- opportunity to refresh and renew workspace image;
- sustainability impact;
- greater autonomy and choice for employees;
- showcasing mobility and new ways of working.

Mobile working allows Nokia employees to meet the requirements of the business on their own terms. One of the key benefits of mobile working is the opportunity to establish a work–life balance.

Nokia Work Style Categories

Deciding which Nokia employees would benefit most from mobile working required a clear classification of worker needs. Nokia derived three different classifications based on each employee's mobility patterns. The three categories were: mobile, campus mobile and desk-based.

- **Mobile worker.** One of the main determining factors of a mobile worker is that they spend about a third or more of their time out of the office environment. When working outside the office the mobile worker may work between many different locations. When working within the office environment, the mobile worker will have no dedicated workspace allocated, and they will choose a workspace that is appropriate to their immediate work priorities. Most of the mobile worker's daily work resources are portable or can be accessed remotely.

- **Campus mobile worker.** The campus mobile worker typically works in one location, spending more than two-thirds of their time in the office but with only about 20% of their time at a desk when they are in the office. With only 20% desk utilization, campus mobile workers are not provided with a dedicated workspace. The campus mobile worker tends to work away from their desk, either in meetings, informal meetings with colleagues or even working in laboratories. Campus mobile workers use mobile technologies, which give them the flexibility to be connected remotely.
- **Desk-based worker.** In contrast to the mobile and the campus mobile workers, the desk-based worker tends to be located in a specific place. This means they spend approximately 80% of their time in the office and at a desk when they are in the office. Most of their daily work resources (e.g. people, technology, equipment, and documents) are and should be office-based. A dedicated workspace is typically used but should be cleared and made available for others to use when out of the office.

Nokia Employee Benefits of Mobile Working

With the adoption of mobile working comes the *'non-territorial'* workplace. This means that whilst employees give up a designated desk they are offered a range of different types of space to use as and when appropriate. The Nokia House project has identified the benefits that employees can obtain from working in the new spaces. The benefits identified are:

- **Work–Life balance: 'Be where you want to be'.** More choice in workspaces to support different work needs and privacy. The workspaces should include: more drop-in workspaces, phone rooms, project spaces, meeting rooms and quiet/study rooms.
- **Cutting edge design with more fun and improved ergonomics.** The Nokia House design includes fewer solid walls and more glass for transparency. Lively colours were used throughout the working environment along with new furniture to change the atmosphere significantly. Design of mobile work areas should include: WLAN everywhere, desks that are height-adjustable, easily adjustable ergonomic chairs, flat screens, ergonomic keyboard and mouse and lockable storage/filing cabinets near the workstation area.

- **Perfect connections.** Better technical support in all work spaces. This should include: access to recharging facilities for devices, easy printing facilities everywhere, high quality (HALO) conference rooms in 30 global sites and wireless access for employees, visitors and guests.

Working Floor Design Concept

Nokia believes that the workplace is formed of many different parts and spaces. This led to the workplace being fitted with a range of different types of space and employees having the flexibility to choose the space most appropriate for their immediate task. One of the working floor concepts was to create neighbourhoods around which employees could relate. The range of different workspaces provided at Nokia House can be seen in Figure C3.3.

Workplace of Choice

The working floor design concept aims to enhance communication and employee transparency. The design principle adopted was that by striving to have a more open working environment a transparent way of working could be supported. The team should be free to shape group space, moving away from individual personalization.

Several spaces are provided for informal interaction leading to a more collaborative and vibrant workplace. Upholstered, soft wall rooms and semi-open circular

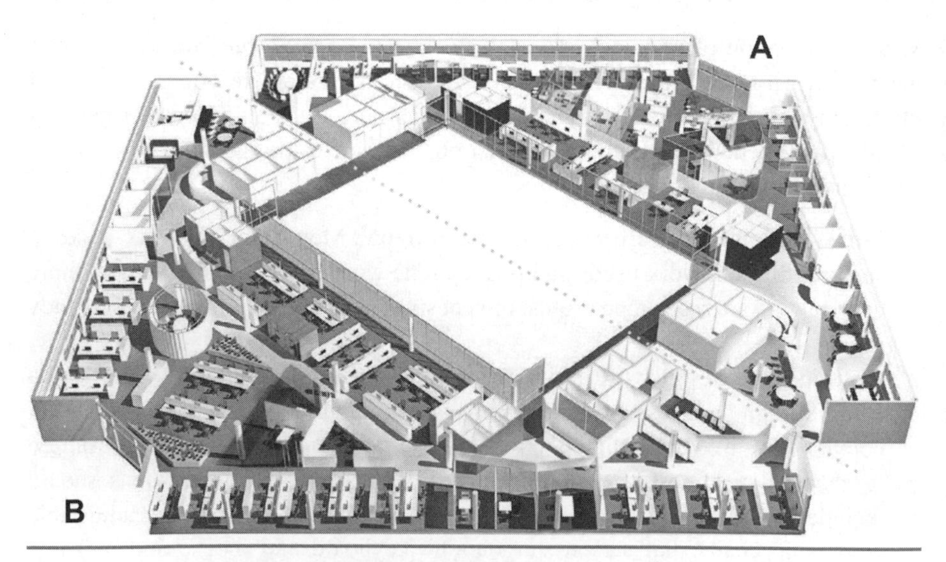

Figure C3.3 Nokia House working floor design concept. *(Reproduced with kind permission of Clive Wilkinson Architects)*

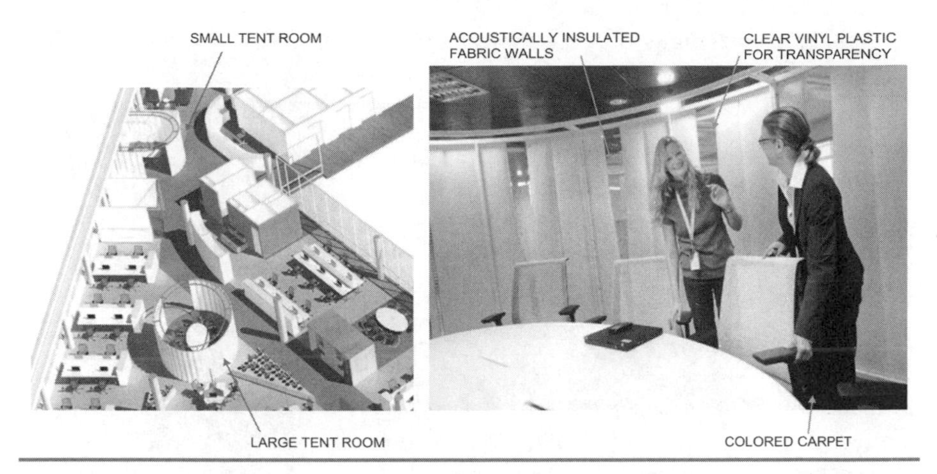

Figure C3.4 Upholstered soft wall rooms. *(Reproduced with kind permission of Clive Wilkinson Architects)*

meeting rooms which are made of acoustic fabric can be used for impromptu meetings or people looking to work alone. An example of these types of spaces can be seen in Figure C3.4.

Whilst Nokia encourages interaction and collaboration with other team members, it is also acknowledged that there are times when privacy is required for private phone conversations or quiet, concentrated work. Therefore Nokia have provided phone booths within the floor design. An example of the phone booth used at Nokia can be seen in Figure C3.5.

Change Management

Nokia's experience of implementing and moving to new ways of working identified that each Nokia employee underwent a certain amount of personal change. It is important to acknowledge that individuals react to change in different ways. If changes are imposed on the workforce by someone else, such as changes in office layout and office relocations, then more time needs to be given for employees to accept and embrace the change.

Implementing a major change requires preparing the leadership/management team as well as the employees. One of the greatest predictors of a successful new working environment is having the local management team support the concept and engage the employees in the decision making process.

A summary of the lessons learnt by Nokia can be summarized as follows:

- **Clear understanding of objectives.** The local management team should have a clear understanding of the output that is to be achieved by changing the working

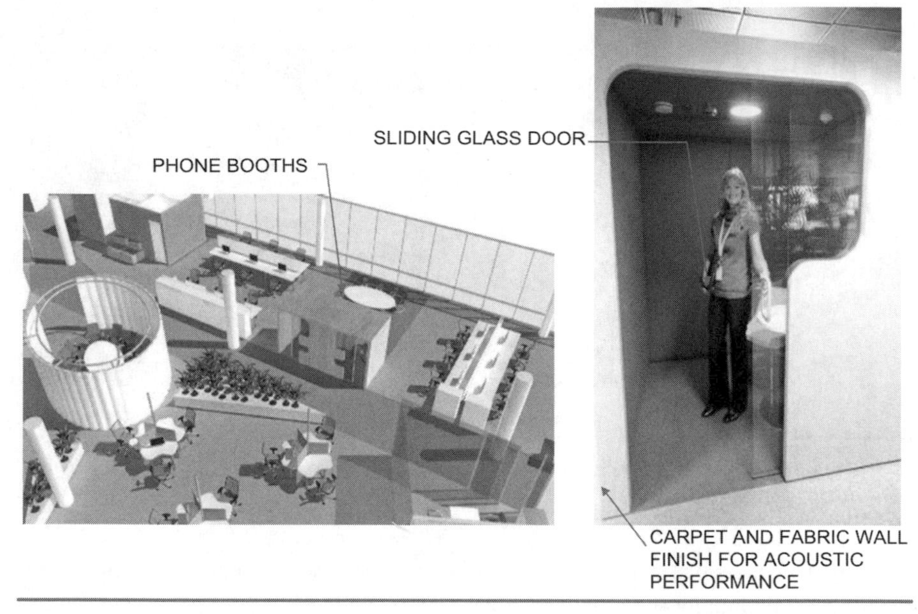

Figure C3.5 Phone booth in Nokia House. *(Reproduced with kind permission of Clive Wilkinson Architects)*

environment. They should all also be provided with the tools that enable them to achieve the desired output, such as communication packages and agreed protocols.

- **Change agents.** It is important in any change management process that *'project champions'* are identified. These project champions will be the change agents for Nokia and therefore it is important that they believe in and are committed to the output of the change management process. They play a pivotal role in convincing others of the benefits of the new work environment, and therefore they need to walk the talk.
- **Visible support from management.** The local management team need to have high visibility during the change management process. This visibility means that they can lead by example by being the first to adopt the new ways of working.
- **Communication.** Keeping people informed and updated of the changes that are being undertaken is an essential part of the change management process. Therefore presentations to employees should be pitched at the appropriate level. To ensure the message is communicated to all employees, many different ways of communicating should be adopted. These could include Internet, videos, email, and coffee talks. Communication is not a one-way process and therefore

opportunities for two-way communication and feedback need to be integrated into the communication strategy. Communication is so important during the change process that the communication and change management team need to be an integral part of the project organization.

- **Fun.** As much as people desire straightforward information in times of change, it is also important not to be too serious. Injecting a bit of fun into the change management process will make the process easier. Some suggestions might include adopting cartoon characters to deliver messages in written or video communications, or, as Nokia did, creating videos about extreme office behaviour.
- **Celebrate.** Once the new working environment and the change management process are complete, then it is important to celebrate your successes. Simple events like move-in breakfasts can help to welcome employees and show off achievements.

Conclusion

The introduction of mobile working requires an integrated solution involving both real estate and human resource management. Nokia's real estate department reports directly to the human resources department. The human resources department was an important project stakeholder in the Nokia house refurbishment project. Workplace projects bring a unique opportunity (and sometimes challenge) to balance employees needs with cost efficiency.

One of the major considerations of a workplace project is the location of the head office building. The positioning of a head office building near to the appropriate labour pool ensures that staff with the relevant qualifications can be recruited. Nokia's head office in Espoo is ideally located to a labour pool that meets its business needs.

Having established a location near to an appropriate labour pool the next consideration is to design a working environment (design/technology/services) that attracts and retains key talented people. In addition, the workplace can be used to communicate to the employees Nokia's brand identity and corporate values.

Nokia are committed to creating a sustainable work environment. This starts in the construction phase with the use of sustainable materials, and embraces all aspects of the physical environment including energy efficiency, space efficiency and services such as recycling in catering facilities, etc. In fact, Nokia has made a commitment that all new construction projects will be LEED certified to at least a Gold standard. Nokia believes that embracing sustainable practices in the office will help lead to good behaviour for employees at home and throughout their lives, and this is also an important factor in attracting key talent.

Workplaces are constantly changing due to the new technological developments such as virtual communication tools. In addition, employee expectations increasingly include demand for a better work–life balance. Nokia's experience has identified that the new generation of office workers is comfortable with collaboration without face-to-face meetings. Since teams working on product development are hardly ever all located in the same building, Nokia uses communication tools to enable virtual collaboration.

However, Nokia is mindful that it needs to be careful not to design offices only for the future generations. The organization acknowledges that it has a multi-generational workforce and considerations should also be given to all generational needs. Issues to be considered will include: work–life balance, less hierarchical structures, design features and service provision. At the same time, Nokia recognizes that work processes shape the environmental needs for its employees, so personal and generational preferences need to be balanced with functional requirements.

Implementing a new work environment can have an impact on the behaviour of its occupants. However, it is Nokia's view that the new mobile workplace concept is only playing catch-up with human behaviour. There has long been a disconnection between current working practices and traditional workplace planning principles. Typically, employees expect a traditional work environment, even though their working practices have changed. The challenge is to align these, and to manage the change. Having accepted that people are already working in a mobile way, Nokia decided to create a workspace that supported mobile working.

A major change management lesson learnt by Nokia was that when altering an employee's work environment it is important that communication starts as soon as possible. People need time to digest, understand and adapt to the change. Also, provision needs to be made for employees to have input into the change process.

Nokia's mobile workplace concept allows for a range of different work styles. Each work style is supported by the appropriate technology and workspace allocation. By adopting the mobile workplace concept, Nokia achieves the cost benefits of reduced head office space, whilst at the same time enhanced productivity is achieved by giving employees the autonomy and flexibility to choose the most appropriate working environment that best meets their needs.

Case Study 4

Headquarter Reconfiguration: The Hong Kong Jockey Club

Danny S.S. Then
The Hong Kong Polytechnic University

About the Hong Kong Jockey Club

The Hong Kong Jockey Club (HKJC) is one of the largest racing organizations in the world. Horse racing is the most popular spectator sport in Hong Kong and through its subsidiaries; the Club is the only authorized operator of horse racing. The Hong Kong Jockey Club was founded in 1884 and changed from an amateur to a professional organization in 1971.

The HKJC is a company limited by guarantee with no shareholders and obtains its net earnings from racing and betting. The Club is the largest single taxpayer in Hong Kong, contributing around 6.8% of all taxes collected in 2008/09 by the government's Inland Revenue Department. A unique feature of the Club, much admired worldwide, is its not-for-profit business model whereby its surplus goes to charity and community projects. Over the past decade, the Club has donated an average of one billion Hong Kong dollars every year to hundreds of charities and community projects. Today, the Club ranks alongside organizations such as the Rockefeller Foundation as one of the biggest charity donors in the world.

The vision of the HKJC is:

> *To be a world leader in the provision of horse racing, sporting and betting entertainment, and Hong Kong's premier charity and community benefactor.*

And its mission is:

> *To provide total customer satisfaction through meeting the expectations of all Club customers and stakeholders – the racing and betting public; lottery players; Club Members; charities and community organizations; Government; and ultimately, the people of Hong Kong – and thereby be one of Hong Kong's most respected organizations.*

Corporate Real Estate Asset Management. ISBN: 978-0-7282-0573-4

The core values of the Club are reflected in its corporate motto of:

'One club, one team, one vision' driven by commitments towards customers, community, people, teamwork, quality and business sustainability and innovations.

The Club is directed by a 12-strong Board of Stewards, headed by a Chairman, who provide their services gratis. Day-to-day management is conducted by a Board of Management of eight Executive Directors, headed by the Club's Chief Executive Officer.

Property Portfolio and the Project Site

The Club's property portfolio in Hong Kong Special Administrative Region (SAR) comprises of two racecourse sites (Happy Valley and Shatin) and staff housing covering a total site area of almost 775,000 m^2, with just over 381,000 m^2 in covered built-up area. The total gross floor area of the portfolio is just less than 900,000 m^2.

The project site is the headquarters building (HQ) of the Club, located on Hong Kong island. The HQ building, constructed in 1995, has 17 floors with a basement level, housing the corporate executives and heads of functional departments and their staff. The HQ building is located within the Happy Valley site, one of the two racecourses owned and managed by the Club's Property Department (Figure C4.1).

The Property Department reports to the Executive Director for Membership Services and Property. The Property Department management team comprise of the Head of Property with four functional managers with responsibilities for Estate, Planning & Design, Operation & Maintenance, and Contracts Management. The Property Department's mission is:

To provide reliable and cost effective property services of the highest standard with an aim to deliver total customer satisfaction in the following areas: Estate, Planning & Design and Operations & Maintenance.

The project details are contained in Table C4.1.

Business Drivers for the Reconfiguration/Restacking Project

The underlying motivation for the project was the recognition of the need for change to cater for the growth in headcount and to upgrade the building infrastructure after almost 15 years of continuous operation, driven primarily by the Club's motto of *'One club, One team, One vision'* and reflected through a corporate aspiration for

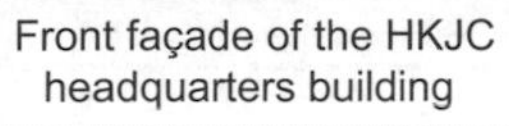

Front façade of the HKJC headquarters building

Aerial view of the HQ building, showing position adjacent to the racecourse

Figure C4.1 Hong Kong Jockey Club headquarters building

more teamwork, more collaboration, less boundary and less hierarchy. This is effectively workplace transformation, which is explored in Chapter 6.

A survey of current occupancy patterns and interviews with key stakeholders, as part of a commissioned Master Planning Study which included benchmarking against best practice and similar buildings in Hong Kong, confirmed opportunities for more effective space utilization together with the introduction of sustainability features. The Master Planning Study recommended different space standards and restacking options for the HQ building.

The outcome was a decision to conduct a pilot project covering the complete reconfiguration of the 12th Floor (12F) as proof of concept and benefits, with an option to extend the same for other floors in the future.

TABLE C4.1 Hong Kong Jockey Club Project Details

Project	Reconfiguration/restacking of offices at the Hong Kong Jockey Club Headquarters
Client	Hong Kong Jockey Club
Design and build contractor	M. Moser Associates Limited
Furniture supplier	Lamex

The key principles that responded to the Club's business needs were encapsulated in the following objectives for the 12F reconfiguration project:

1. accommodate business growth and long term flexibility;
2. more efficient use of space;
3. improve communication;
4. improve brand and image;
5. help attract and retain top talent;
6. address IT and M&E Infrastructure needs;
7. sustainable + environmental friendly solutions;
8. health and safety requirements.

The above project objectives will be discussed under three integrated headings, taking into consideration the measures taken and the benefits achieved.

Accommodate Business Growth and Long Term Flexibility Through More Efficient Use of Space

Background

M. Moser Associates Ltd was commissioned to conduct a Master Planning Study of the HQ Building in early 2008. The study included interviews of key users from each of the executive divisions and departments, a survey of existing use of the space with the HQ Building, comparative market analysis of space standards of similar buildings in Hong Kong and current best practices in planning approach.

The Master Planning Study conducted provided the following findings:

1. The interviews of stakeholders at the HQ revealed a need to improve the work environment and to cater for business growth. However, sentiments were also expressed of the need to handle the human aspects of the potential change management with care. Some samples of comments are listed below.
   - 'Change is required to improve the work environment and accommodate business growth.'
   - 'It is not easy to find a meeting room – support centralized meeting floor for large and external meetings.'
   - 'Some believe in open office concept but need to consider the privacy requirements of Executives and HODs.'
2. The survey of existing use of space confirmed that there was scope for rationalization of space use through restacking of most of the existing floors to promote better teamwork and collaboration.

3. The findings considered that a total of slightly over 3500 square metres of existing uses of space, spread over a number of floors, can be further optimized by reclaiming current spaces that are either vacated, underutilized or unused; reviewing spaces occupied by filing, and duplication of reception areas.
4. The comparative market analysis of existing space standards of similar buildings use revealed a number of mismatches in terms of overall space density, number of space standards, ratio of enclosed office to open workstation, amount and size of shared meeting rooms, and amount of open/shared space. Space standards are explored more fully in Chapter 5. A summary of some comparative data are shown in the Table C4.2.

Solutions Development

Based on the above findings and through discussions with the key stakeholders, two sets of principles were agreed that would form the basis for solutions development for the pilot project covering the 12F of the HQ building. These are summarized in Table C4.3.

Using the principles in Table C4.3 as a guide, several options were evaluated using revised space standards for staff at different levels. The outcome of the evaluation produced a layout for the 12F, with a gross floor area of around 1230 m^2, with the capacity to seat 104 staff comfortably.

The agreed layout resulted in an average density of 9.6 m^2 per seat, comprising of a combination of enclosed and open-plan individual workspaces and enclosed and open-plan meeting spaces. The consultation process between key stakeholders from the client and consultant team captured key design aspirations that were incorporated

TABLE C4.2 Summary of Comparative Space Standards Data

Market	Parameter	Existing HKJC HQ
105–120 sq ft	Seat density (area per seat)	122 sq ft (1–16F)
2–3	Number of space standards	9
2–4%	% of private room	10%
96–98%	% of open workstation	90%
Yes	Centralized location for meeting space	No
Yes	Shared meeting room	Some
8–10%	Open/social space	1%
75% (4–6 persons)	Meeting room size	75% (more than 8 persons)

TABLE C4.3 Principles for Space Planning and Developing Stack Plan

Principles for space planning	Principles for developing stack plan
Best practice planning recommends: ■ 3 to 4 space standards ■ provide social space ■ share all meeting facilities Centralized location for large meeting rooms No multiple reception space	Efficiency of space: ■ keep as many people as possible in the HQ ■ provide long term flexibility Minimize business disruption ■ double moves ■ minimize project duration Cost avoidance ■ infrastructure investment Risk avoidance IT and critical system disruption Working relationships ■ support interaction and collaboration

as key design parameters for the reconfiguration of the 12F layout. Key design parameters for the internal reconfiguration included:

1. increased natural daylight and general openness in design scheme;
2. new simplified space standards with a modular approach;
3. incorporation of sustainable and environmental features in the choice of material and equipment.

Key features of the revised configuration for the 12F include:

- private rooms: 13
- open seats: 91
- enclosed meeting seats: 23
- open meeting seats: 19

TABLE C4.4 Comparison of Status Quo with Agreed Option

	Status quo	Agreed option
Total seats accommodated (BF to 17F)	1,487 seats	1500–1600 seats
Density (sq ft per seat)	122 (average)	111 (12F only)
Meeting room seats	346 seats	420–480 (incr. by 21–39%)
Cost of churn	\$A + 35%	\$A + 15%
Long term flexibility	N/A	85% Modular + Flexible
Shared view and daylight	20% staff has access to daylight	50% staff has direct access to daylight
Health and safety issue	45 seats located on Basement Floor	Relocate staff out of Basement Floor (BF)

Table C4.4 shows a comparison of the results of the agreed option against the current status quo.

The proposed Stacking Plan has recommended a shared centralized meeting floor to be located at the 1st Floor, together with the relocation of all workstations from the Basement Floor. A major consideration is to maximize the number of workspaces with direct access to natural daylight.

Address IT and M&E Infrastructure Needs by Adopting Sustainable and Environmental Friendly Solutions that Meet Statutory Health and Safety Requirements

The reconfiguration of the 12F of the HQ Building also provided opportunities for incorporating the latest technological innovations from a sustainable viewpoint. A number of 'green' technical features with intelligent controls were implemented as part of the reconfiguration of the space planning layout.

- T5 tubes, electronic dimmable ballasts and NanoFlex coated luminaries
- LED downlights along corridors and non-workspace areas
- Digital Addressable Lighting Interface (DALI) lighting control system
- Direct Digital Control (DDC) for fan coil units
- Fresh air demand ventilation control
- Solar films coating for perimeter glazed surfaces for solar radiation control
- Comprehensive metering for performance measurement

All the features are aimed at improving the status quo from the perspectives of reducing energy consumption and water usage: improve indoor air quality, more user friendly controls for the user's individual workspace environment, and provide better monitoring and control of overall energy and water consumption in the HQ building. Targets set for energy consumption and IAQ standards for the project are aimed at meeting or exceeding current standards.

A summary of the key benefits of the green technical features will now be explored.

Energy Efficient Features of New Lighting System

The upgrading of lighting systems with a view of incorporating the latest energy saving features is a major consideration of the technical specifications for the pilot project (Figure C4.2).

- The reconfigured floor incorporated a new T5 reflecting lighting system with electronic dimmable ballasts to control the LUX output. New lighting fixtures featured NanoFlex coating, which can achieve the same lighting level with

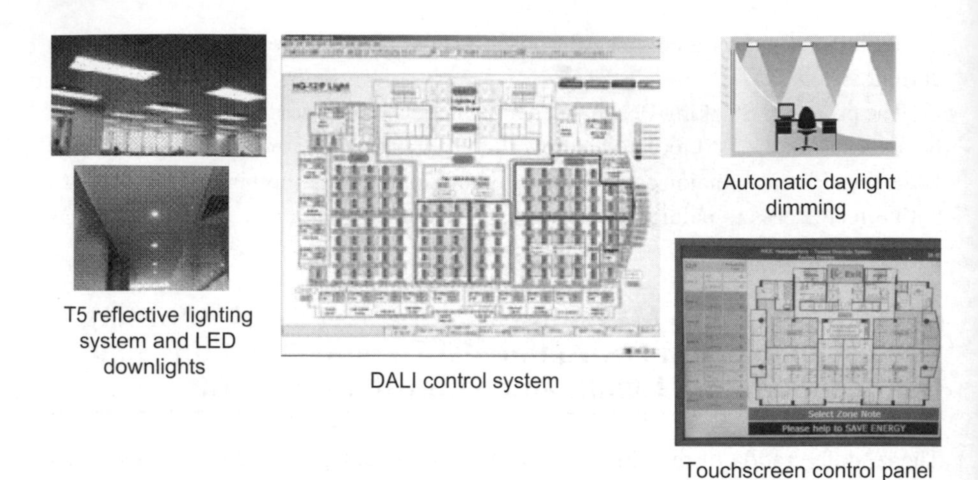

Figure C4.2 Upgraded lighting systems

lower wattage. Results from BMS monitoring have demonstrated a total energy saving of 31.4% compared with pre-renovation consumption figures.

- LED downlights were installed along corridors and non-workstation areas including pantry, toilets and cargo lift lobby. LED downlights are more energy efficient than traditional halogen downlights and have a much longer life cycle.
- A key feature of the upgraded lighting system in terms of control features is the incorporation of the Digital Addressable Lighting Interface (DALI) lighting control system. The system has highly flexible control features to change the lighting control zones to suit future office and seating layout. Additional features include: automatic daylight dependent dimming control in perimeter offices; different dimming control modes during lunch hour, motion sensors inside enclosed offices, and separate settings for ad hoc requests during extended office hours.

Energy Efficient Features of Air Conditioning System

The upgrading to the air conditioning system for the 12F renovation comprised of two elements, the Direct Digital Control (DDC) Fan Coil Unit System and the Demand Ventilation Control.

Open plan office fan coil units were controlled by the DDC system. Parameters such as room temperature, supply air flow and on/off schedule were controlled by the central BMS system via DDC controllers to provide a comfort environment and also achieve energy saving.

VAV Unit and CO_2 sensors were installed to control the fresh air supply to 12F. The VAV Unit will receive signals from the CO_2 sensors and control the modulating damper to regulate the fresh air supply. CO_2 contents within the premises will be controlled below 1000 ppm. This design will save more energy by automatically adjusting the fresh air supply relative to the number of occupants in the premises.

Both these systems greatly enhance the control capabilities of the air conditioning provision to the occupiers of space through a finer level of zoning controls, automatic sensors and temperature settings for normal and ad hoc requests during extended office hours.

Enhancement to Building Management Systems

This upgrade provides additional capabilities in the monitoring of energy utilization for the renovated floor. In order to prove that the pilot project delivers what it sets out to achieve, the ability to measure energy inputs from various sources is critical in terms of performance measurement against set targets. Power analysers were installed to closely monitor the electrical consumption of lightings, sockets and fan coil units. An energy meter was installed to closely monitor the air conditioning load demand within the premises. The upgrade will ensure that all energy consumption is captured, measured and trended for analysis and performance reporting on a routine basis. Continuous monitoring of these parameters helps to setup a good operation program and is also vital for proactive maintenance.

Solar Films on External Glazed Areas for Solar Radiation Control

Solar film was installed on curtain wall windows to reduce the solar heat gain and UV penetration which saves energy for the air conditioning load; and also provide a comfortable working environment. The ultra safety and security features of the film also helped in holding the windows in place during hurricanes, severe winds and other extreme conditions.

Help Attract and Retain Top Talent through Improved Corporate Brand and Image

The Club is one of the largest employers in Hong Kong, with some 5300 full-time and 21,000 part-time staff. The provision of appropriately designed and functional workspace is clearly critical, not only in terms of meeting health and safety requirements, but is becoming more critical in recruiting and retaining top talent from the marketplace. The growing awareness of the need to be a socially responsible corporate citizen to potential employees is clearly part of the motivation to incorporate the latest technological advances in the upgrading of the 15-year-old building infrastructure and redesign of the current workspace at the HQ building.

Project Management

The project planning and duration for the 12F reconfiguration of the HQ building is shown in Figure C4.3. The project duration from confirmation of layout to handover was 18 weeks, with on-site construction taking 9 weeks.

Before designing the new workspace, the Strategic Planning Team from M. Moser Associates conducted comprehensive research and analysis as part of the Master Planning Study. The research involved a series of interviews, questionnaires and feedback that both engaged staff in the change process and sought to establish what they wanted and needed from the new office design. The analysis provided much needed information relating to work patterns and departmental adjacencies that could be addressed through or supported by the strategic restacking planning for the whole HQ building.

The research and analysis process played a vital role in both encouraging buy-in from key staff members and initiating dialogue between departments that would need to collaborate more extensively in future.

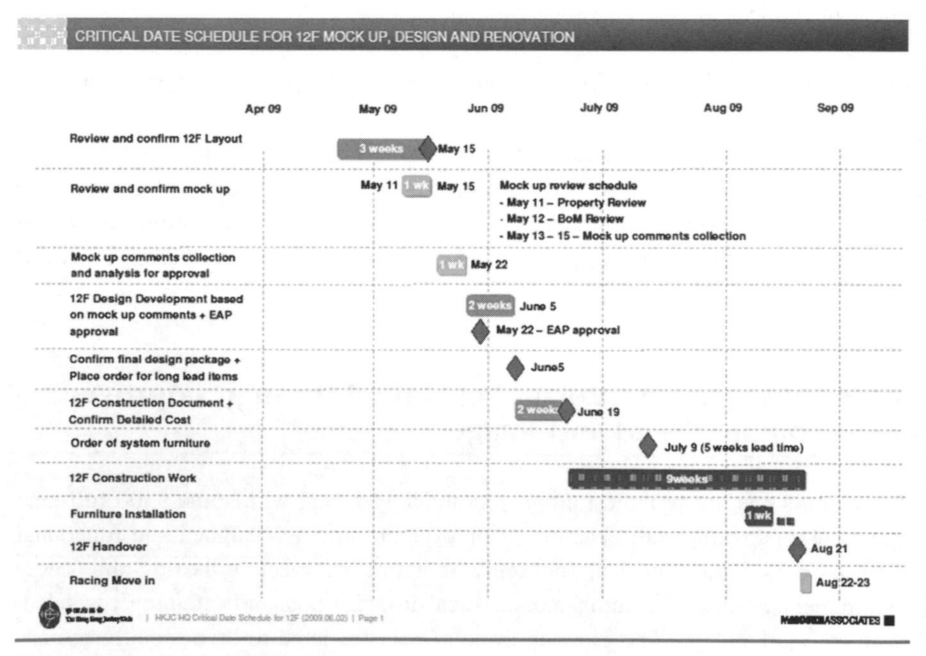

Figure C4.3 Project plan

View of reconfigured open-plan workspace

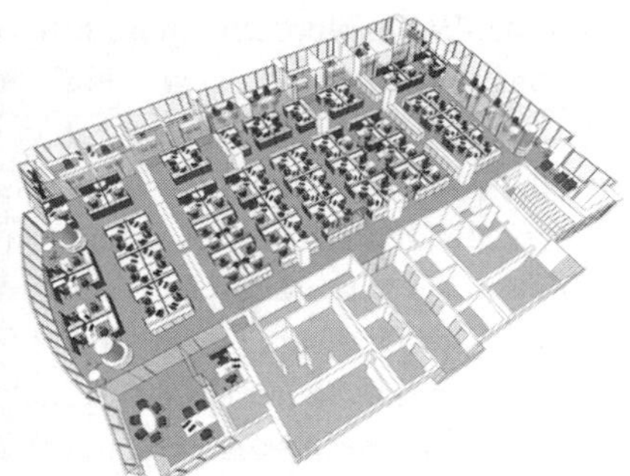

Generic layout of reconfigured 12F

Figure C4.4 Reconfigured open-plan workplace

Project Outcomes

Preliminary assessment of the outcomes of the pilot project of the reconfiguration of the 12th Floor of the HQ building has been positive. As proof of concept, the project has demonstrated that, with proper planning and adoption of industry best practice norms, a more intensive use of valuable space with improved flexibility and efficiency is feasible without loss of amenities. The revised layout has resulted in an almost doubling of capacity with design features that encourage better collaboration and communication between staff and departments (Figure C4.4). The incorporation of green technical features has not only resulted in a visually pleasing, but also a more environmentally friendly workplace environment that optimizes external daylighting and indoor air quality through intelligent controls.

The HKJC is currently considering options for the restacking of the other floors of the HQ building together with the benefits resulting from the reconfiguration of the 12F pilot project. Future reconfiguration and restacking will probably be carried out in phases in order to cause minimum disruption to normal business operations.

Corporate reconfiguration and restacking are major change projects that will affect most employees. One of the lessons learnt is the importance of communication, not only between all stakeholders within the project team, but more importantly, the need to inform and consult all staff that will be affected at an early stage.

Conclusion

Corporate relocation, reconfiguration and restacking projects are responses to changes in organizational demand for a business support resource, i.e. functional space; as a result of business dynamics and changes in corporate strategy and direction. We explore this more fully in Chapter 7. Such projects represent actions that are needed to adjust the building supply in terms of realigning workspace capacity and standards to reflect contemporary needs (especially technological advances and sustainability demands) to support modern ways of working. In successfully managing this change, a balance has to be struck between potential benefits to the corporation and impact on the individuals working for the corporation.

This approach validates the need for a holistic approach to CREAM. It demonstrates that strategic interventions such as a headquarter reconfiguration demands an integration of the themes explored throughout this book.

Case Study 5

Sustainability: Amazon Court, Prague – A Win–Win Solution

About Amazon Court, Prague

Prague may not seem the obvious choice for a case study on sustainability as there is a lack of BREEAM and LEED assessments; and a slightly less mature development market than other European nations. What makes Amazon Court special and worthy of inclusion is:

- the attitude of the developer, Europolis, who retain and manage their own developments;
- the innovation of the design and sustainability components;
- most significantly, a real and proven approach to lower occupancy costs and added value which we believe creates a true 'win–win' situation for both landlord and tenant.

In addition, the authors have access to detailed data; on a comparative study of total occupancy costs carried out by Petr Žalský, Leasing & Marketing Manager at Europolis, as part of an MSc Consultancy Project supervised by the authors at Sheffield Hallam University.

Europolis Real Estate Asset Management

Established in 1990, Europolis is involved in the management of the investment activities of Europolis AG, EBRD, AXA Real Estate Investment Managers and Union Investment Real Estate in the Central and Eastern European (CEE) region. The company focuses on acquisition, development and asset management of offices, shopping centres and distribution facilities. The core values of the company include professionalism, excellence in construction and management as well as transparency in business. Europolis is currently managing a portfolio of more than 40 projects in Austria, the Czech Republic, Hungary, Poland, Croatia, Romania, Russia and Ukraine.

Corporate Real Estate Asset Management. ISBN: 978-0-7282-0573-4

Europolis is an interesting company in that, unusually for a property company, it develops, asset manages and retains many of its projects and it states in its home page that:

We believe it is our tenants who provide the added value …

Examining this statement not only sets out a clear picture of innovation and customer focus but a genuine and unique positioning in the marketplace which ensures a long term perspective in managing their assets and creating a win–win relationship between property owner, asset manager and tenant. It is these relationships and perspective that we believe allow the case study we present here to be both unique and special.

The Building

Amazon Court is a landmark office building located in the Prague district of Karlín (see Figure C5.1). It was completed and taken over from the general contractor in June 2009 and is the third building of the River City Prague (RCP) development by Europolis, following on from the successful Danube House and Nile House projects. Its location adjoins the Vltava River and has good access to two Metro stops, tram and bus networks.

Figure C5.1 Amazon Court, main elevation

TABLE C5.1 Amazon Court Project Team Composition

Developer	Europolis Real Estate Asset Management s.r.o
Landlord	RCP Amazon s.r.o
Project manager	Arcadis Project Management s.r.o
Architects	Schmidt Hammer Lassen
Architectural and engineering	Atrea spol s.r.o
Engineers	RFR Paris
General contractor	Metrostav
Sustainability engineers	ZEF Sustainability Consultants & Low Energy Engineers

Designed by the Copenhagen based architects Schmidt Hammer Lassen, the building comprises seven floors, above ground, providing 19,800 m^2 of office accommodation; 2200 m^2 of retail space; and three levels of parking below. Amazon Court was the winner of the Business Centres category in the MIPIM Architectural Review *Future Project* Award 2008 and the *Best Office Development* 2009 in the Construction Investment Journal Awards 2009.

The project team for Amazon Court is set out in Table C5.1.

The building utilizes a number of sustainable and environmentally friendly concepts, and is designed to use nearly half the amount of energy of a comparable conventional building whilst providing superior internal climate comfort. Local services provided within the complex reduce the need to travel to the city centre and promote the well-being of the tenants, whilst specialized landscaping complements the building and its surroundings.

A number of international and local companies have chosen this neighbourhood for their headquarters including KPMG, Allianz, Bovis, Deloitte, Procter & Gamble, Ford, Kraft Foods, Arthur D. Little, Pioneer Investment, Motorola, Nokia and Unilever.

Urban Sustainability Components

Location

The Amazon Court building, as part of RCP, provides office premises and service orientated retail premises on the ground floor. These include a restaurant and cafeteria. Other services like post office, ATM and convenience store can be found in the adjacent Danube House and Nile House. The ground floor atriums are designed as public spaces; and the public realm is enhanced through revitalization of the river embankment and surrounding areas. The closest mass transport stations are within 3–7 minutes' walking distance. The RCP project is also directly incorporated into a newly established system of cycling pathways.

Access

RCP is connected to the existing Karlín urban area by a footbridge over Rohanske Nabrezi Street. The footbridge was built by the developer together with the first building of RCP. A new traffic light controlled junction was incorporated to improve pedestrian safety and manage the traffic flow. Public transport is easily accessible, with two metro stations of different lines (Florenc and Krizikova) within 5–7 minutes' walk and a number of trams just 3 minutes' walk from the building.

Building Design

The entire building is constructed of reinforced concrete with flat slabs, including the concrete perimeter walls. The concrete structure is exposed to the internal premises to provide the energy system with sufficient thermal mass and improve tightness of the building. The external façade is made from natural stone. Almost every module has been provided with an openable window, allowing natural ventilation, thereby reducing the need for air conditioning. Sun-shading, mainly on the interior, is provided on all façades (see Figure C5.2) except the north elevation and ground floor.

Figure C5.2 Photograph showing internal shading and ETFE roof to the atrium

The atrium roof is made of clear ethyl tetrafluoroethylene foil (ETFE), the same material used in the 'Bird's Nest' stadium in Beijing, which complements the natural ventilation system and replicates the aural characteristics of the city.

Planning and Implementation Process

The overall RCP project development incorporated in its design new infrastructure, including an upgrade of vehicular access, a pedestrian bridge, cycling paths, an underground road, comprehensive sustainable landscaping features and improvement of flood protection to a 1000-year event period.

Environmental Sustainability Components

Energy Efficiency, Climate Control and Air Distribution

The internal climate control system was elected on the basis of the following key features:

- low energy consumption;
- low maintenance requirements;
- flexibility and easy adaptability;
- healthier, smooth-running and comfortable working environment.

The system used in Nile House, Danube House and Amazon Court is an underfloor displacement ventilation system. The floor displacement principle is that the air is supplied into the raised floor at each level, entering the occupied zones from floor grilles to form a low level lake effect of fresh treated air. The supplied air rises in a plume from around the heat sources, e.g. the occupants, office machines and windows. The floor level grilles and their frequency throughout the floor plate are shown in Figure C5.3. As the warm air rises it collects pollutants and unwanted heat from the exposed lighting and computer equipment.

The air is drawn from the riverside via the air inlet turrets. This air then enters the underground tunnels constructed in heavy reinforced concrete. The air is then drawn through filters and across heat exchangers fed by borehole cooling water, or heat pump cooled water at peak summer conditions, cooling the incoming air by up to 16 degrees in the summer and pre-heating it by up to 25 degrees in the winter. The air is channelled to the air handling units and then split to supply each zone of building via separate supply routes.

The operational principles are different for summer, mid-season and winter periods. The air is being changed up to four times per hour depending on weather season (traditional air condition system can change the air only approximately

Figure C5.3 Floor grilles in an unoccupied floor

1.4 times per hour). The photographs in Figure C5.4 show the elements of the air distribution system.

Reduction of Emissions

The low energy approach unique to the development saves 35–50% of energy compared to conventional air conditioned buildings while providing the same or better internal climate control. This approach helps to reduce not only the user's operating cost but also the emissions connected to production and consumption of energy.

The combination of the following factors is expected to reduce the CO_2 emission by at least 50%:

- solar passive design (southern exposed atrium);
- shading;
- selected glazing;
- the thermal mass exposure (raised floor, no false ceiling and exposed concrete perimeters walls);
- night ventilation providing 'free of charge' required air temperature level for building cooling;
- earth tubes (concrete air supply tunnels);
- the heat recovery within the air handling units;
- building management system and inverter controls;

The air inlet turrets adjoining the river from which clean air is drawn

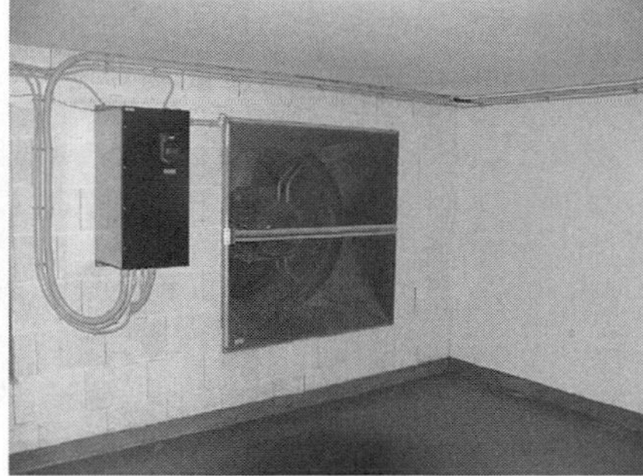

The very large fans pulling air into the basement of the building

The filtering, cooling and distribution equipment

The air exhausts and illustration of the increased ceiling height as suspended ceilings are not required

Figure C5.4 Photographs showing the journey of the fresh air into the building

Renewable Energy Resources

Ground source heat pumps and earth tubes are utilizing renewable energy of the earth and providing a stable temperature throughout the year.

Environmentally Friendly Measures Used in the Building

A waste management system is integrated into the building design. All waste is separated for recycling.

All the chemicals used for cleaning in the building are free of hazardous materials and are thinned with water which allows for discharge into the sewerage system. The company providing cleaning service is required to hold an adequate certificate (e.g. ISO 14001).

The lighting used in the building is low energy consumption.

Economic Sustainability Components

Rental and Utilization Rate

On introduction to the market the rents varied from €16.90 to €17.90 per m^2 per month (rents are quoted on a monthly basis in Central and Eastern Europe). This price level is above the average for a similar building in a comparable location, according to data provided by Jones Lang LaSalle, with prices ranging from €14.75 to €15.50 per m^2. As will be demonstrated in the research presented below, the higher rent is offset by other savings due to the energy efficiencies inherent in the design. The previous buildings in RCP, Danube House and Nile House are both 100% leased, none of the tenants has used their 'free of penalty break option' after year 5 (in Danube House), indicating high levels of satisfaction.

Building Operational Performance

The low energy consumption performance system saves 35–50% of energy compared to conventional air conditioned buildings while providing better internal climate comfort (100% fresh air, better daylighting and better views). The electricity consumption in Danube House and Nile House was 85 kWh/m^2/year for office space in the year 2008; similar figures are also expected for Amazon Court. This number represents all electricity consumption needed for building operation (cooling, lighting of common areas, lifts etc.), excluding the tenants' own electricity consumption used for IT, table lamps, kitchenettes etc.

Floor Space Efficiency

The width of the two office floor plates is 18.3 m, with the typical layout being a triple-bay layout with 7.3 m window-to-core distance. This plan width offers flexibility and efficiency for various office layouts (open plan and individual office layout). The planning module for office accommodation is 1.5 m. The building is designed to have a population density of 10 m^2 perimeter workspace zone (HNF) per person. The clear floor to ceiling height in the office premises is 3 m.

The design follows the BCO (British Council for Offices) best practice wherever it was not in conflict with local statutory regulations. The floor area in the building was measured on the basis of the Directive for Calculation of the Leased Floor Area of Office Premises (*Richtlinie zur Berechnung der Büromietfläche*), also called GIF.

UTILIZATION AND FLEXIBILITY OF THE PROPERTY

The property is designed to provide maximum comfort and minimum disturbance during operation. Four different solutions for vertical communication are provided allowing flexibility from whole floor plate occupation down to areas of only 250 m^2. Personal and freight lifts connect the basement loading bay and office floors and provide the office users with regular office supply/delivery, space alternations material delivery, moving furniture etc., which ensures minimum disturbance to tenants.

The air distribution system both provides flexibility and maximizes adaptability within the building. The system allows very easy and low cost internal layout modifications with minor changes usually involving only the manual adjustment of the floor air inlets, as shown in Figure C5.5.

Research indicates that alterations to the space planning of a floorplate, due to churn or restructuring in a conventional office building, will involve:

- significantly higher costs (15–30%) and
- significantly higher operation interruption time (downtime) (30–40%)

than in Amazon Court. This is because reconfiguration may involve significant adjustments to air conditioning and other systems.

Financial Analysis of the Total Occupancy Costs

Research supervised by the authors has been undertaken which proves the viability of the environmental approach adopted at Amazon Court and demonstrates how it

Figure C5.5 Photograph of air distribution grille; simply turning the grille adjusts the air flow

breaks the *'vicious circle of blame'*, creating a win–win situation for developers, landlords and tenants.

The research examines the total occupancy costs over a five-year occupation period, including rent, rent inducements, service charge costs and the initial fit out costs. The study compares a 'non green' control building of similar size, floorplate, specification and location, with Amazon Court.

The basic data used in the research study are set out in Figure C5.6.

In addition to the significant savings in the service charge, the fit out costs for Amazon Court are significantly less because of the designed-in flexibility and adaptability of the cooling system and internal layout.

When all of the total costs are analysed in a five-year cash flow the following outcomes are achieved as set out in Figure C5.7.

The Win–Win Outcomes

For the developer and fund the sustainable solutions are more expensive to create initially but are offset by an increased rental and potential to attract and retain customers due to the integrated sustainability approach. The lack of tenants in the two previous projects initiating their break clauses indicates strong satisfaction and 100% tenant retention in the two adjoining buildings.

Marketing is focused on appropriate communication of the total occupancy costs. A strong argument can be delivered (now proven) that the total costs of occupation will be significantly lower than occupying a non-green building, in a similar location. In addition to the tangible costs savings, the flexibility, adaptability and air quality offer significant 'intangible' benefits to the occupier (as discussed in Chapter 6), including:

- the ability to open windows and control fresh air ventilation;
- the improved flexibility created by the ground floor air flow ventilation and simple do-it-yourself control of air flow allowing greater densities than standard air conditioning systems and highly improved flexibility;
- lower costs of churn and restructuring due to the above flexibility of design;
- ability to increase density of occupation without expensive reconfiguration of the air management system;
- an air management system that constantly replaces air with new fresh air rather than dilutes it with re-circulated air and the consequent increase in staff alertness and potentially productivity (see Figure C5.8);
- increased ceiling height and well-being factor of spaciousness and clean design.

Further research is planned by the authors to validate the intangible benefits and measure the increased productivity expected in Amazon Court compared to a traditional building.

Rental data and service charge data for Amazon Court

AREAS	LEASED AREAS	RENT	INITIAL RENT	INITIAL RENT
OFFICES	m^2	RATE EUR/m^2/month	EUR p.m.	EUR p.a.
NRA1 excl. terraces	*1375.00*	*17.40*	*23,925*	*287,100*
Terraces external	0.00	0.00	0	0
Terraces internal	0.00	0.00	0	0
Terraces total	*0.00*		**0**	**0**
NRA1 incl. terraces	1375.00		23,925	287,100
NRA2	*20.62*	*17.40*	*359*	*4,308*
NRA1+NRA2 total excl. terraces	1395.62		24,284	291,408
NRA1+NRA2 total incl. terraces	1395.62		24,284	291,408
STORAGE	*0.00*	*0.00*	*0*	*0*
PARKING	*0*	*0.00*	*0*	*0*
ADD-ON	1.50%	*TOTAL RENT*	*24,284*	*291,408*

SC	SERVICE CHARGES	SERVICE CHARGES
RATE EUR/m^2	EUR p.m.	EUR p.a.
3.50	*4,813*	*57,756*
0.00	0	0
0.00	0	0
	0	**0**
	4,813	57,756
3.50	*72*	*864*
	4,885	58,620
	4,885	58,620
2.59	*0*	*0*
0.00	*0*	*0*
TOTAL SC	*4,885*	*58,620*

Rental and service charge data for 'non green' control building

AREAS	LEASED AREAS	RENT	INITIAL RENT	INITIAL RENT
OFFICES	m^2	RATE EUR/m^2/month	EUR p.m.	EUR p.a.
NRA1 excl. terraces	*1375.00*	*14.50*	*19,938*	*239,256*
Terraces external	0.00	0.00	0	0
Terraces internal	0.00	0.00	0	0
Terraces total	**0.00**		**0**	**0**
NRA1 incl. terraces	1375.00		19,938	239,526
NRA2	*139.42*	*14.50*	*2,022*	*24,264*
NRA1+NRA2 total excl. terraces	1514.42		21,960	263,520
NRA1+NRA2 total incl. terraces	1514.42		21,960	263,520
STORAGE	*0.00*	*0.00*	*0*	*0*
PARKING	*0*	*0.00*	*0*	*0*
ADD-ON	10.14%	*TOTAL RENT*	*21,960*	*263,520*

SC	SERVICE CHARGES	SERVICE CHARGES
RATE EUR/m^2	EUR p.m.	EUR p.a.
4.50	*6,188*	*74,256*
0.00	0	0
0.00	0	0
	0	**0**
	6,188	74,256
4.50	*627*	*7,524*
	6,815	81,780
	6,815	81,780
2.59	*0*	*0*
0.00	*0*	*0*
TOTAL SC	*6,815*	*81,780*

Figure C5.6 Rental and service charge data for both buildings

Amazon Court:

ANALYSIS		
TOTAL OCCUPATIONAL COST		
RENT	1,311,336	EUR
SERVICE CHARGES	302,870	EUR
FIT-OUT	174,763	EUR
TOTAL	**1,788,969**	**EUR**

Non-green building

ANALYSIS		
TOTAL OCCUPATIONAL COST		
RENT	1,185,840	EUR
SERVICE CHARGES	422,530	EUR
FIT-OUT	232,059	EUR
TOTAL	**1,840,429**	**EUR**

Notes:
- **rent free of 6 months applied for both buildings**
- **service charges are paid for 62 months (2 months time for fit out included)**

Figure C5.7 Five-year cash flow analysis of total occupancy costs of both buildings

A traditional dilution air conditioning system

The displacement ventilation system used at Amazon Court

Figure C5.8 Illustration of the underfloor displacement principle compared to a traditional air conditioning system and its potential impact on productivity

Index

25 years institutional lease (UK), 85
Abbey National, 79–80
Academy realm and organizations, 184
Acceptance and change, 197
Access:
  Amazon Court case study, Prague, 280
  main railway stations, 192
Actium Consult case study:
  change management, 245–6
  Corporate Real Estate Asset Management, 251
  description, 243
  guide to better use of office space, 249–51
  mission, 243–4
  occupier solutions, 244–5
  psychology, 251
  Total Occupancy Cost Survey, 117, 246–9, 251
Adding value:
  brand identity, 50–2
  CREM, 45–6
  human assets, 49–50
  increasing productivity, 48–9
Ageing population, 9–10
Agora realm and, organizations, 184
Air conditioning and Hong Kong Jockey Club, 272
Alterations (leases), 96
Amazon Court case study, Prague, 277–88
  access, 280
  building design, 278–9, 280–1
  building operational performance, 284
  economic sustainability, 284–5
  emission reduction, 282
  environmental sustainability, 281–3
  Europolis Real Estate Asset Management, 277–8
  financial analysis, 285–6
  floor space, 284–5
  planning/implementation, 281
  renewable energy, 283
  rental, 284, 287
  total occupancy costs, 285–6, 288
  urban sustainability, 279–81
  utilization/flexibility of property, 285
  win–win outcomes, 286–7
Ancillary space and NUA, 188
Andersen, Arthur, 117
'Apples and pears' (benchmarking):
  capitalization, 115
  measurement, 112–14
  occupation, 114–15
Arlington Business Services, 86, 102–3
'Arlington Way', 102
Arundel Building, Sheffield, 217
Arup Sustainable Project Appraisal Routine, 210
Asda Store, Bootle, Merseyside (UK), 214
Asset management, 31–2
Assignment and subletting (leases), 95–6
Australia:
  green leases, 224–6
  Green star sustainability programme, 207

BAA:
  benchmarking, 102
  case study, 89
  tenant retention, 99–100
Baby boomers, 7
Balanced scorecard tool, 136–8
Bazaar scenario, 23–4
BBC, 80–1, 82
BCO *see* British Council for Offices
Beard, J. & Roper, K., 202, 218
Becker, F., 163
Becker, F. & Stelle, F., 156–8, 166
Behaviour and physical environment, 156–7
Behavioural environment and office productivity, 155–69
Belshazzar's feast scenario, 26
Ben & Jerry's, 203
Benchmarking:
  'apples and pears', 112–15
  CREAM strategy, 134–6, 138, 139
  criticism, 115–18
  current situation, 117–18
  definition, 107–9
  holistic real estate, 131–4
  IPD, 121–6, 126–30
  Jones Lang LaSalle, 130–4
  measurement, 110–11
  Office of Government Commerce, 121–6
  practice, 110–12
  space standards, 118–21
  take up in real estate industry, 116–17

Benchmarking Real Estate Excellence, 233
Bibby, Howard, 86, 89, 102
Bitner, M.J., 156
Body Shop, 203
BOMA *see* Building Owners and Managers Association
Bon, R., 34
Boots the Chemist, 114
BOSTI Associates, 162
Brand and identity, 180–1
Brand identity and adding value, 50–2
BREEAM *see* Building Research Establishment Environmental Assessment Method
Brenner, P. & Cornell, P., 157–8
Briefing:
  process, 187–8
  strategy, 188
Brill, M. et al, 162–4
British Council for Offices (BCO), 284
British Institute of Facilities Management (BIFM), 32
British Standards:
  BS7750 (environmental management), 209
  BS EN ISO 1401 (1996), (environmental management), 209, 215, 220
British Standards Institute (BSI), 32
Brown, Lester R., 201
Brundtland Commission, 201–3
Bruntwood case study, 102–3, 138
  benchmarking, 107, 238–42
  customer focused approach, 89
  customer service, 235–7, 238–42
  description, 233
  green leases, 237
  publications, 236
  Red Rooms initiative, 237–8
  success, 234–8
  values, 238
Budgets and funding, 178
'*Buildings ecology*', 167
Building Owners and Managers Association (BOMA), 113–14
Building Research Establishment Environmental Assessment Method (BREEAM):
  Asda Store, Bootle, Merseyside, 214
  assessment, 212–14
  description, 211
  Digital Realty Trust, 214
  green buildings, 229
  'in use' system, 214–15
  sustainability, 207, 208
Buildings:
  Amazon Court, Prague design, 280–1
  BREEAM excellent status, 213
  management systems and HJKC, 273
  operational performance in Amazon Court, Prague, 284
  sustainability, 205, 210–11
  workplace supply, 188–95
Burolandschaft, 143
Business:
  adjacencies, 179–80
  direction, 175–6
Business Link, UK, 202

CABE *see* Commission for Architecture and Built Environment
California Building Sustainable Taskforce, 218
Calvert, John, 97
Campus mobile workers, 259
Capitalization and 'apples and pears', 115
Carbon Reduction Commitment (CRC):
  Energy Efficiency Scheme, 237
  UK, 227, 230
Carbon reduction and sustainability, 227–8
CASBEE *see* Comprehensive Assessment System for Building Environment Efficiency
CEE *see* Central Eastern European
Celebrations at work, 263
CEN (European Committee for Standardization), 32
Central Eastern European (CEE) region, 277
Certification and sustainability, 210–11
Change:
  driving forces, 22
  leases, 96
  management, 5, 196–7
  unthinkable, 27–8
Change management:
  Actium Consult case study, 244–5
  Nokia, 2613
Chartered Institute of Public Finance and Accountancy, 117
Climate Change Act (2008), UK, 227
Clive Wilkinson architects, 253, 260
Clutter in offices, 187
Code for Leasing Business Premises, 237
Collaboration space in offices, 148–9, 182
Collaborative/individual work environments, 164–6
Comfort in offices, 160–1
Commercial Lease Code, 85

Commercial property and sustainability, 207
Commission for Architecture and Built Environment (CABE), 180, 187
'*Common space*', 157
Commons and caves concept, 144
Company accounts:
 leases, 75–8
 sale-and-leaseback, 71–2
Competitive benchmarking, 109
Competitor analysis, 16–19
Comprehensive Assessment System for Building Environment Efficiency (CASBEE), 210–11, 213
Concentration space in offices, 148, 182
Concentration v. communication, 157–8
Conlan, Tara, 82
Connectivity (theoretical framework), 168–9
Contemplation space in offices, 149, 182
CoreNet Global, 1, 8, 208
Cornell Balanced Real Estate Assessment (COBRA) model, 49
Corporate business and sustainability, 204–5
Corporate DNA, 57
Corporate PFI, 80–1, 82
Corporate Real Estate Asset Management (CREAM):
 Actium Consult, 251
 adding value, 46
 behavioural office environment, 170
 benchmarking, 134–6, 138, 139
 budgets and funding, 178
 building/workplace supply, 189
 business direction, 175–6
 change, 1
 collapsing of boundaries, 31–6
 competitor analysis, 16–19
 Corporate Social Responsibility, 228
 corporate strategy, 36–9, 38
 cultural aspirations, 177–8
 cultural web, 42–4
 headcount/departmental breakdown, 178–9
 holistic approach, 276
 management, 12–13
 office productivity, 169
 organizational structure, 176–7
 PESTEL analysis, 13
 power of buyers, 18
 power of suppliers, 17–18
 providers, 17
 psychology, 251
 services, 176–7
 skills/education, 34–6
 strategic alignment model, 52–4, 195–6, 198
 strategic solutions, 31
 Web 2.0 technologies, 153
 workstyle analysis, 181
Corporate Real Estate Management (CREM):
 adding value, 45–6
 description, 34
 ICT, 44–5
Corporate real estate and sustainability, 208–11
Corporate relocation:
 briefing, 187–8
 building and workplace supply, 188–95
 introduction, 173
 optimum alignment, 195–7
 organizational/business demand, 175–87
 strategic space planning alignment process, 173–5
Corporate Social Responsibility (CSR):
 business benefits, 203–6
 report, 201
 sustainability, 202–3, 205–6, 220–1, 230
Corporate strategy and real estate strategy, 39–42
Cost and real estate, 111
CRC *see* Carbon Reduction Commitment
CRE management/strategy, 2
CREAM *see* Corporate Real Estate Asset Management
Creating the Productive Workplace (report), 155
Creativesheffield model, 192–4
CREM *see* Corporate Real Estate Management
CSR *see* Corporate Social Responsibility
Cullen, Mary, 11
Cultural aspirations, 177–8
Cultural web, 42–4
Cushman & Wakefield European Cities Monitor, 190
Customer focused procurement options:
 expectations, 85–9
 finance, 89–92
 introduction, 85
 leasing business premises, 94–7
 reduction of lease length, 92–3
 RICS service charge code, 97–9
 UK strategies, 99–104
Customer focused/traditional real estate management comparison, 90–1
Customer relationship management (CRM), 151
Customers:
 definition, 100
 flexibility, 235
 needs, 100–1
 service (Bruntwood), 235–7, 238–42

DALI *see* Digital Addressable Lighting Interface
Dantesque scenario, 25
Data gathering techniques, 186
DDC *see* Direct Digital Control
De Jonge, H. et al, 120
Definition:
benchmarking, 107–9
customer, 100
performance measurement, 107
sustainability, 201–3
DEGW architectural practice, 146, 175, 185
Demand and supply, 174
Demand to supply deviation, 194–5
Denial and change, 197
Desk-based workers, 259
Detailed briefs, 187
Digital Addressable Lighting Interface (DALI), 271–2
Digital Realty Trust and BREEAM, 214
Direct Digital Control (DDC) and HJKC, 271–2
Discovery and change, 197
Distraction v. interaction (office), 158–9, 161
Distraction-free work, 187
Distributed workplace, 146–7
DIT *see* Dublin Institute of Technology
Dow Jones Sustainability World Index (DJSI World), 225
Drucker, Peter, 145
DTZ Global Office Occupancy Costs Survey, 107–8
Dublin Institute of Technology (DTI), 20
Duffy, Frank, 143

EBS *see* Environment-Behaviour Studies
Eco-Management and Audit Scheme (EMAS) in Europe, 209
Eco-Office (workplace scenario), 26
Economic sustainability in Amazon Court, Prague, 284–5
Edington, Gordon, 99–101
Efficiency and real estate, 111
EFTEethyl tetrafluoroethylene
Emissions reductions in Amazon Court, Prague, 282
Empryean scenario, 26
EMSenvironmental management system
Energy of Buildings Directive (EBPD) in UK, 208
Energy Performance Certificates (EPCs), 208, 215–17, 230
Environment and organizations, 37
Environment-Behaviour Studies (EBS), 166–7
Environmental management system (EMS), 209–10
Environmental sustainability in Amazon Court case study, Prague, 281–3
Environmentally friendly products, 204
EPC *see* Energy Performance Certificate
EPF *see* European Property Foundation
Ergonomics and Nokia offices, 259
Ethyl tetrafluoroethylene (ETFE) foil, 281
EU *see* European Union
European Cities Monitor, 190
European Property Foundation (EPF), 94
European Regional Development Fund (ERDF), 221
European Union (EU):
directives, 208, 215–16, 222
Emissions Trading Scheme, 227
Exploring Corporate Strategy, 36
External benchmarking, 110
Externally mobile workstyle, 185

Facilities management (FM), 2–4, 32–3, 80, 86, 153, 241
Facility Management Association of Australia (FMA), 33
FASB *see* Financial Accounting Standards Board
Fasset Management, 103–4
Finance:
Amazon Court, Prague, 285–6
assets, 34
real estate procurement, 62–4
Financial Accounting Standards Board (FASB), 74
Fit factor and NUA, 188
Five Forces analysis, 16–17
Fleming, D., 167
Floor Rentable Area (FRA), 113–14
Floor space in Amazon Court, Prague, 284–5
FMfacilities management
'*FM Six Pack*', 114
Food and offices, 187
Ford, Henry and model T production, 142
FRA *see* Floor Rentable Area
Freehold:
leasehold comparison, 57–62
leasehold conversion, 69–73
Fun at work, 263
Functional benchmarking, 109
Future scenarios, 19–20

Gates, Bill, 27–8
Gattaca (workplace scenario), 26

GEA *see* Gross External Area
Gearing ratios, 78–9
General Services Association (US), 124
Generation:
X, 7, 207
Y, 6–9, 28, 207, 218, 251
Z, 207
Generic benchmarking, 109
Gensler workplace surveys, 148, 150, 154–5, 182–3, 204
GIA *see* Gross Internal Area
Gibbon, Paul, 214
Gibson, V., 145–6
Globalization, 3–5
Great Depression, 143
Green buildings, 205–6, 208, 229
Green leases:
Australia, 224–6
Bruntwood, 237
certificate, 226
sustainability, 221–6
United Kingdom, 222–3
Green Property Alliance, UK, 230
Green Star (sustainability programme) in Australia/New Zealand, 207
Gross External Area (GEA), 113
Gross Internal Area (GIA), 113
GSA *see* General Services Association
Gucci Group N.V., 72–3
Guild realm and organizations, 184
Gullsten and Inkinen (Nokia office), 253

Handy, Charles, 145
Harvard Business Review, 78
Hawken, Paul, 201–3
'Hawthorne Experiments', 142–3
Headcount and departmental breakdown, 178–9
Heerwagen, J.H. et al, 164–5
Hermes lease, 223
High Performance Property Initiative, 117
Hilton Hotels, 73
Hive (workplace scenario), 26
HKJC *see* Hong Kong Jockey Club
Holistic real estate and benchmarking, 131–4
Homeworking, 151
Hong Kong Jockey Club (HKJC) case study:
air conditioning, 272–3
building management systems, 273
business growth, 268–71
conclusions, 275–6
corporate branding, 273–4
description, 265–6
flexibility, 268–71
health and safety, 271–3
IT infrastructure, 271–3
Master Planning Study, 267–8
new lighting, 271–2
project management, 274–5
property portfolio, 266
reconfiguration/restacking project, 266–7
solar films, 273
space utilization, 268–71
top talent, 273–4
Hot Property: Getting the Best from Local Authority Assets,, 117
Hotels and wireless Internet, 152
Hughes, J.E. & Simoneaux, B., 11–12
Human assets and productivity, 49–50
'Human Relations' movement, 143

IASB *see* International Accounting Standards Board
IFMA *see* International Facility Management Association
Impact and uncertainty, 22–3
Information and communication technologies (ICT):
Corporate Real Estate, 44–5
Corporate Real Estate Asset Management, 5
distributed workplace, 146
mobile working, 38
real estate strategy, 44–5
services, 181
'*Informal interaction points*', 165
Insurance (leases), 97
'Intelligent Building' concept, 44
Interaction v. distraction (office), 158–9, 161
Internal benchmarking, 110
'*Internal rents*', 114
Internally mobile workstyle, 185
International Accounting Standards Board (IASB), 74, 78
International benchmarking, 110
International Facility Management Association (IFMA), 2, 4
International standards, 209–10
International Standards Organization (ISO):
EN ISO 14001 (2004), 209, 215, 220
environmental management, 209–10
International Total Occupancy Cost Code (ITOCC), 126–30
Investment Property Forum, UK, 219
IPD (investment property databank):
benchmarking, 121–6
corporate occupiers, 133
index, 108

IPD (investment property databank): (*Continued*)
International Total Occupancy Cost Code, 126-30
Occupier model, 115, 135
Occupiers index, 246
Occupiers Standards for office space: (OGC report), 118
portfolios, 116
publications, 92–3, 99
space code, 117
space standards, 120
ISO *see* International Standards Organization
ITOCC *see* International Total Occupancy Cost Code

Jazz scenario, 24
Johnson Controls Incorporated (JCI), 1, 6, 26
Johnson, G. & Scholes, K., 108
Jones, K & White, A.D., 32
Jones Lang LaSalle, 74, 99, 113, 130–4, 191

Kaplan, R. & Norton, D., 136
Knowledge work, 184
'Knowledge worker', 145

Land Securities:
Bruntwood comparison, 237
Landflex, 57, 66–8, 249
Trillium, 80–1, 82
Landlord and Tenant Act1954, 95
Landlords (UK):
history, 86
satisfaction, 87–8
tenant relationship, 85, 89
Landmarks of Tomorrow, 145
Langstone Technology Park, 104
Laptop computers and transportation, 151
'Laptops in the park', 151
Leadership, empowerment, training and communication, 101
Leadership in Energy and Environment (LEED) design in US, 207, 229, 263
Learning space (office), 150
Leasehold:
freehold comparison, 58–61
lease length, 64–5
Leases:
alterations and change of use, 96
assignment and subletting, 95–6
business premises, 94–7
company accounts, 75–8
Green, 221–6
Hermes, 223
insurance, 97
length, 64–5
length of term, break clauses and renewal rights, 95
negotiations, 94–5
on-going management, 97
rent deposit and guarantees, 94
rent reviews, 95
repairs, 96
service charges, 96
LED (light emitting diode) lights, 216, 230, 272
LEED *see* Leadership in Energy and Environment
Length of term, break clause and renewal rights, 95
Lighting and Hong Kong Jockey Club, 271–2
Line manager's role at Nokia, 255–6
Lodge realm and organizations, 184
Lords of Misrule scenario, 23

M&S *see* Marks & Spencer
McHugh, John, 233
Main railway station access, 192
Maire, J., 108
Making ICT Work for You (report), 44
Maperley property company and outsourcing, 79–80
Marks & Spencer (M&S), 58, 220–1
Mayo, Elton, 142
MBA programmes and sustainability, 204
Measurement:
'apples and pears', 112–14
benchmarking, 110–11
success and benchmarking, 101–2
Media Guardian, 82
Mission:
Actium Consult, 243–4
organizations, 101
statements, 176
Mobile technologies, 152
Mobile workers, 258–9
Multi-generational headquarters, 10–12
Multi-generational workforce, 4–5, 5–9
MWB virtual offices, 57, 67, 134
Myerson, J. & Ross, P.184

Nanoflex coatings (lighting), 271
Natural resources management (NRM) systems, 205
Needs (customer), 100–1
Net Internal Area (NIA), 113
Net Usable Area (NUA), 188

New working practices, 2
New Zealand and Green Star (sustainability programme), 207
Newport, UK redevelopment, 221
NIA *see* Net Internal Area
Noise in offices, 186–7
Nokia case study:
  change management, 261–2
  changes in work, 254–6
  conclusions, 263–4
  description, 253–4
  employee benefits of mobile working, 259–60
  Helsinki office, 254–5, 263
  line manager's role, 255–6
  mobile workplace concept, 39, 185, 257–8
  office design, 259
  performance measurement, 255
  space allocation, 256
  target state, 256–7
  team members, 256
  workstyle, 258–9
  working floor design, 260
  workplace of choice, 260–1
Nokia House, Helsinki, Finland, 253–5
Norwich Union, 81–2
Nourse, H., 118

Occupiers Satisfaction Index (OSI), 240
Occupation and 'apple and pears', 114–15
Occupier perspective, 156, 166–8
Occupier solutions, Actium Consult case study, 244–5
Occupiers Standards for Office Space: (OGC report), 118
O' Connor, Donal, 12
Office:
  clutter, 187
  comfort, 160
  distraction, 160
  food, 187
  interaction, 160
  layout, 160–1
  noise, 167
  Nokia concept, 254–5
  open plan, 144
  open plan v. cellular, 162–4
  productivity, 154, 160–1
  security, 187
  workplace, 147–50
'*Office ecology*', 167
Office of Government Commerce (OGC), 117, 121–6, 135, 175, 185
'Office landscape', 143
'*Office occupier*', 155
Office space:
  Actium Consult guide, 249–51
  collaboration, 148–9, 182
  concentration, 148, 182
  contemplation, 149, 182
  learning, 150
  socialization, 150
OGC *see* Office of Government Commerce
Oglesby, Mike, 233
Olson, J., 158–9
On-going management and leases, 97
'*One club, One team, One vision*' motto, 266
Open plan offices, 144
Open plan v. cellular offices, 162–4
Operational assets, 34
'*Organizational ecology*', 144
'*Organizational glue*', 169
Organizations:
  Academy, 184
  Agora, 184
  business environment, 37
  competencies, 37
  corporate relocation, 175–87
  culture, 37, 42–4
  direction/scope, 37
  environment, 37
  expectations/purpose, 37
  Guild, 184
  Lodge, 184
  mission, 101
  resources, 36–7
  structure, 177
OSCAR Programme, 130–4
Osgood, T., 135, 137
OSI *see* Occupiers Satisfaction Index
Out of office working, 151–2
Outline briefs, 187
Outsourcing, 2, 79–80
Oxford Properties Group (USA), 102
OXYGENZ project, 6–8

'Pay as you go' concept, 145
People and buildings interface, 44–5
'*People centred*' approach, 156
Performance:
  benchmarking, 109
  management and real estate, 111
  measurement, 112, 255
  workplace, 153–69
Personal computers (PCs), 144
'*Personal harbour workplace*', 157

PEST (Political, Economic, Social and Technological) analysis, 13, 35
PESTEL (political, Economic, Social, Technological, Environmental, Legislative)
analysis, 13, 28, 176
Peterson T.O.& Beard, J. W., 165–6
Physical assets, 34
Physical environment and behaviour, 156–7
Piecework concept, 143
Policy opinions, 26–7
Porter, Michael, 16
Porter, Richard, 74
Post-occupancy evaluation (POE), 197
Premises and Facilities Management, 86
PricewaterhouseCoopers (PwC), 10–12, 202
Principia ethica scenario, 26
Pringle Brandon architects, 251
Procedures and protocols, 186–7
Processes:
benchmarking, 109
improvement and information management, 101
Procter, Andrew, 243–4
Productivity and workplace, 153–69
Property in Business: A Waste of Space (report), 118, 153
Property management, 32
Public spaces and work, 151
PwC *see* PricewaterhouseCoopers

Quality and real estate, 111

'*Ramp-up*' (concentration), 159
Rank Xerox, 108–9
Rapoport, A., 167
RCP *see* River Court Prague
'*Readiness to change*', 197
Real estate:
management traditional/customer comparison, 90–1
performance management, 111
Real Estate Norm (REN), 191–2
Real estate procurement:
alternatives to freehold/leasehold, 79–83
balance, 64–9
finance, 62–4
freehold to leasehold conversion, 69–72
freehold v. leasehold comparison, 57–62
future options, 73–9
trends, 5
Real estate strategy:
corporate strategy, 39–42
information and communication technologies, 44–5
organizational culture, 42–4
REAL SERVICE organization, 239
Red Rooms initiative, 237–8
Regus virtual offices, 5, 68–9
Reichheld, Fred, 240
REN *see* Real Estate Norm
Renewable energy in Amazon Court, Prague, 283
Rent:
Amazon Court, Prague, 284, 287
deposit and guarantees, 94
reviews, 95
Repairs (leases), 96
Residents (workstyle), 185
Resistance and change, 197
River Court Prague (RCP) development, 278–80
ROCE ratio, 78–9
Rockefeller Foundation, 265
Rothlisberger, F.J., 142
Royal Institute of Chartered Surveyors (RICS):
asset management, 31–2
green buildings, 208
green space in US, 28
Making ICT Work for You, 44
Net Internal Area, 113
Property in Business: A Waste of Space, 118, 153
Property Management Awards, 102
service charge code, 97–9, 237
sustainability, 206–7

Sale-and-leaseback option, 69–73
SAS (Scandinavian airline), 109
Savitz, Andy, 202
Scarrett, D., 88
Scenarios:
bazaar, 23–4
Belshazzar's feast, 26
Dantesque, 25
Empryean, 26
Jazz, 24
lords of misrule, 23
planning summary, 27–8
principia ethica, 26
Socratic systems, 24
titans of avarice, 26
wise counsel, 24–5
Schnelle brothers, 143

Schramm, U., 174
'Scientific management', 141
Security in offices, 187
Senge, Peter, 204–5
SENvICS *see* Sustainable Environmental Improvements in the Commercial Sector
Services:
  charges (leases), 96
  CREAM, 176–7
  flexibility, 65–6
  offices, 151
Shamrock Organization, 145
Shareholder value, 134
Sherry, Dom, 26
Siemens:
  Annual Corporate Social Responsibility Report (2007), 205
  Lindenplatz location, 206
  Munich-Perlach offices, 206
  SRE Italy, 206
SIG *see* Sustainability Interest Group
Skills and education, 34–6
Socialization space (office), 150
Socratic systems scenario, 24
Solar films and HJKC, 273
South Central office building, Manchester, UK, 233
Space standards and benchmarking, 118–21
Space to Work, 184
SPeAR *see* Sustainable Project Appraisal Routine
Starbucks, 151
Steele, Fritz, 144
STEP (Sociological, Technological, Economical and Political) analysis, 13–14
Strategic alignment:
  adding values, 45–52
  Corporate Real Estate Management, 34
  corporate strategy, 36–45
  CREAM, 5, 52–4
  integrated approach, 31–4
  skills/education, 34–5
  space planning process, 173–5
Strategic Alignment Benchmarking Component, 135
Strategy:
  analysis, 12–16
  benchmarking, 109
  briefing, 188
  management process, 37–9
  planning, 20–1
Sturge, King, 23–4
Support space and NUA, 188
Surveying Sustainability: A short Guide for the Property Professional, 206
Sustainability:
  buildings, 205
  carbon reduction, 227–8
  CASBEE system, 210
  certification, 210–11
  commercial property, 207
  corporate business thinking, 204–5
  corporate real estate, 208–11
  Corporate Social Responsibility, 205–6, 220–1
  definitions, 201–3
  green leases, 221–6
  intermediate/can be measured aspects, 218
  introduction, 201
  labelled buildings, 210–11
  MBA programmes, 204
  measurement, 216–19
  practice, 219–21
  professional perspective, 206–7
  Royal Institute of Chartered Surveyors, 206–7
  soft/intangible/difficult to measure aspects, 218
  trends, 3–5
Sustainability Interest Group (SIG), 219
Sustainable Environmental Improvements in the Commercial Sector (SEnvICS), 221, 223
Sustainable Project Appraisal Routine (SPeAR), 210
SWOT (strengths, weaknesses, opportunities and threats) analysis, 14–16, 28, 35, 132–3, 176

'*t word*' (tenant), 233
Taylor, Frederick Winslow, 141–3
Technology:
  infrastructure, 2
  requirements, 181
Tenant:
  landlord relationship, 85, 89
  retention and cash flow, 91–2
The Age of Unreason, 145
The Costs and Financial Benefits of Green Buildings, 218
Titans of avarice scenario, 26
TOCS *see* Total Occupancy Cost Survey
Total occupancy costs in Amazon Court, Prague, 285–6, 288
Total Office Cost Survey (TOCS) (Actium Consult), 117, 243, 246–9, 251

Total property outsourcing, 79–80
Traditional v. contemporary property management, 88
Traditional/customer focused real estate management comparison, 90–1
Transportation and laptop computers, 151

United Kingdom (UK):
  BREEAM, 207
  Business Link, 202
  Carbon Reduction Commitment, 227, 230
  Climate Change Act (2008), 227
  Energy of Buildings Directive, 208
  feudal system, 86
  green leases, 22–3
  Green Property Alliance, 230
  Investment Property Forum, 219
  landlords, 86–8
  Office of Government Commerce, 117, 121–6, 135
United States (US):
  Energy Policy Act (2005), 208
  Environmental Protection Agency, 202
  General Services Association, 124
  green space, 228
  Leadership in Energy and Environment (LEED) design, 207
  public estate, 208
University of Reading annual survey of CRE, 1
Unthinkable change, 27–8
Urban sustainability in Amazon Court, Prague, 279–81
Utilization and real estate, 111

Veterans, 7
'*Vicious circle of blame*', 286
Virtual work, 152–3
Vision statements, 176, 187
Voice over Internet Protocol (VoIP), 152–3

Web 2.0 technology, 152
Welcome Workplace study, 11, 149, 182–6
Western Electric Company, 142
Wise Counsels scenario, 24–5
Work:
  celebrations, 263
  fun, 263
  modes, 182–4
  nature, 145–7
  Nokia, 254–5
  public spaces, 151
  Starbucks, 151
  virtual, 152–3
Working beyond Walls - The Government Workplace as an Agent of Change, 175, 185
Workplace:
  collaborative/individual environments, 164–6
  definition, 162
  management trends, 4
  qualities, 163
  scenarios, 25–6
  transformation, 5
Workplace transformation:
  evolution, 141–5
  introduction, 141
  nature of work, 145–7
  Nokia case study, 253–364
  office, 147–50
  out of office, 151–2
  performance/productivity, 153–69
  virtual work, 152–3
Workspace and NUA, 188
Workstyle:
  analysis, 181–6
  externally mobile, 185
  internally mobile, 185
  residents, 185
Work–life balance, 259
Wrigglesworth, P. & Nunnington, N., 189
Wright, Frank Lloyd, 142

Xyratex, 104

Yates, Rob, 208, 233–6, 240–2

Zairi, M., 108
Žalský, Petr, 277